MUSCULAR SYSTEM - ANATOMY, FUNCTIONS AND INJURIES

SARCOPLASMIC RETICULUM

STRUCTURE, DEVELOPMENT, ROLES AND EVOLUTION

MUSCULAR SYSTEM - ANATOMY, FUNCTIONS AND INJURIES

Additional books in this series can be found on Nova's website under the Series tab.

Additional e-books in this series can be found on Nova's website under the e-book tab.

MUSCULAR SYSTEM - ANATOMY, FUNCTIONS AND INJURIES

SARCOPLASMIC RETICULUM

STRUCTURE, DEVELOPMENT, ROLES AND EVOLUTION

RIGERS BAKIU

New York

Library of Congress Cataloging-in-Publication Data

ISBN: 978-1-63482-839-0

Library of Congress Control Number: 2015939086

Published by Nova Science Publishers, Inc. † New York

This book si dedicated to all my family,
my colleagues and my friends
for their great support and patience
during the time of writing this book

Contents

Preface

This book explores the latest research on the sarcoplasmic reticulum (SR) in skeletal, cardiac and smooth muscle. It examines the structure, development and roles of the SR. The SR is present in all vertebrate classes studied to date including fishes, amphibians, reptiles, birds, and mammals, but the contribution of SR calcium release varies considerably among species. Using a comparison approach, the structure, development and roles differences are described from an evolutionary point of view. Medicine and animal sciences master and PhD students are invited to explore the architecture and identify the roles of SR, but also junior or senior scientist can find the recent scientific reports about SR. Obviously the audience is very large and includes people from different university level and scientific areas. The subject areas are physiology, biochemistry, molecular biology, biophysics, molecular evolution and molecular medicine, while the related job functions are represented by physician, biochemist, molecular biologist, veterinarian and biologist.

Chapter 1

Structure of the Sarcoplasmic Reticulum

Abstract

Muscle cells, including cardiomyocytes, contain a specialized smooth endoplasmic reticulum (ER), the sarcoplasmic reticulum. Ca^{2+} release and reuptake by the sarcoplasmic reticulum trigger the contraction and relaxation of the myofibrils. The ER is a multifunctional signaling organelle that controls a wide range of cellular processes such as the entry and release of Ca^{2+}, sterol biosynthesis, apoptosis and the release of arachidonic acid (AA). One of its primary functions is as a source of the Ca^{2+} signals that are released through either inositol 1,4,5-trisphosphate (InsP(3)) or ryanodine receptors (RYRs). Since these receptors are Ca^{2+}-sensitive, the ER functions as an excitable system capable of spreading signals throughout the cell through a process of Ca^{2+}-induced Ca^{2+} release (CICR). This regenerative capacity is particularly important in the control of muscle cells and neurons. Thus, it would be important to present first the structure of the muscles and simplify the complexity of skeletal muscle physiology by fiber type classification, where differences are observed from neuromuscular transmission to release of intracellular Ca^{2+} from the sarcoplasmic reticulum and the resulting recruitment and cycling of cross-bridges [480].

Fiber Type Classification

Muscle fiber type classification schemes have progressed over the years. The original distinction of fiber type was based on color: red (high myoglobin) and white (less myoglobin) correlated with speed of contraction (slow *vs.* fast, respectively) and fatigability (fatigue resistance *vs.* fatigable, respectively). This advanced to different classification schemes: one was based on contractile properties and histologic staining for metabolic enzyme activities - slow oxidative (SO), fast oxidative, glycolytic (FOG) and fast, glycolytic (FG) [1]; and a second based on the pH lability of actomyosin ATPase staining, with fibers classified as type I, IIa, IIb, or IIx. Corresponding to this latter classification scheme, fibers were found to be primarily composed of a single MyHC isoform, and specific antibodies were developed to identify the expression of different MyHC isoforms. Today, fiber type classification is primarily based on immunoreactivity to antibodies, which are specific for different MyHC isoforms. These antibodies can also be used in Western blot analysis to determine the MyHC isoform composition and content of individually dissected muscle fibers.

In general, muscle fibers classified histochemically as type I, IIa, IIb, and IIx are composed of $MyHC_{slow}$, $MyHC_{2A}$, $MyHC_{2B}$, and $MyHC_{2X}$ isoforms, respectively [2]. Type IIb fibers in some muscles are an exception, as these fibers often co-express $MyHC_{2X}$ together with $MyHC_{2B}$ [3, 4].

In addition to the four adult MyHC isoforms, there are also embryonic ($MyHC_{emb}$) and neonatal ($MyHC_{neo}$) MyHC isoformsthat are usually found in skeletal muscle fibers during embryonic and early postnatal development. During early myogenesis, primary myotubes express $MyHC_{slow}$ in combination with the $MyHC_{emb}$ isoform [5]. With the progression from primary to secondary myotubes, expression of $MyHC_{neo}$ isoform appears and continues through the early stage of postnatal development as adult fiber types mature. Although muscle fibers singularly expressing $MyHC_{slow}$ (type I fibers) are present during early postnatal development, singular expression of other MyHC isoforms appears only later in postnatal development [6]. Thus, distinction fiber type classification is limited to more mature muscle fibers Pathophysiological conditions are also associated with MyHC co-expression. For example, MyHC isoform co-expression in single muscle fibers has been reported in conditions of hypothyroidism [7], aging [8], and neurodegenerative diseases such as spinal muscular atrophy [9] and amyotrophic lateral sclerosis [10]. The presence of MyHC isoform coexpression may reflect fiber type conversion, although clear evidence for such conversion is lacking. It is clear however, that whenever MyHC isoform co-expression occurs, it blurs the unambiguous distinction of muscle fiber type.

Motor Unit Types – Influence of Innervation

There is substantial evidence that the central nervous system exerts a predominant influence over fiber type (MyHC isoform expression) in skeletal muscle. This is evident by the homogenous fiber type composition of motor units (Figure 1).

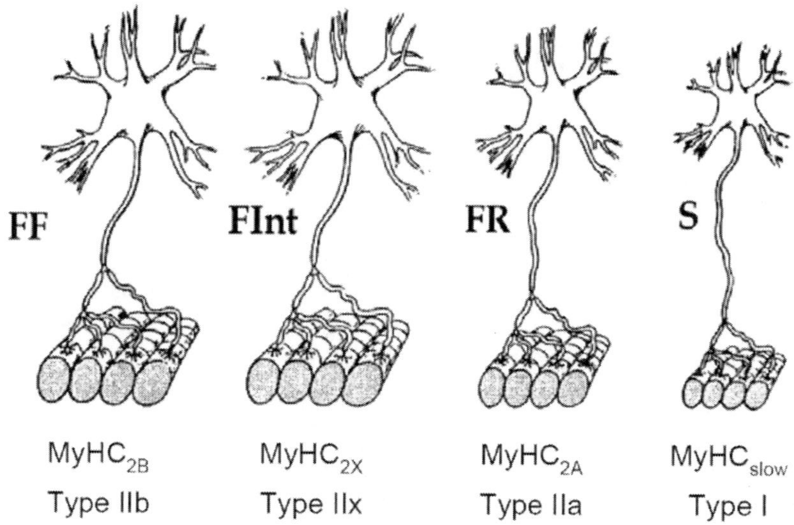

Figure 1. Four motor unit types - slow (type S), fast-twitch fatigue resistant (type FR), fast-twitch fatigue intermediate (type FInt), and fast-twitch fatigable (type FF) - are classified based on the contractile and fatigue properties of the innervated muscle fibers. Each motor unit type comprises a specific fiber that expresses a single myosin heavy chain (MyHC) isoform.

Within a motor unit, a motor neuron action potential activates the group of muscle fibers it innervates in an all-or-none fashion [11-13]. Motor units are classified into slow- (type S) or fast-twitch units, based on the mechanical properties of the motor unit muscle fibers [12]. Fast-twitch motor units are further classified based on their fatigability as fast-twitch fatigue resistant (type FR), fast-twitch fatigue intermediate (type FInt), and fast-twitch fatigable (type FF). This classification of motor unit types parallels the classification of type IIa, IIx and IIb muscle fibers, respectively (Figure 1) [14, 15]. It remains controversial whether activation history or neurotrophic factors are responsible for the predominant influence of innervation on fiber type and MyHC isoform expression. In studies where the diaphragm muscle was inactivated for a prolonged period of time by cervical spinal cord hemisection, fiber type composition and singular MyHC isoform expression in fibers was unaffected [16], [17]. In contrast, when neural influence was removed by denervation, MyHC isoform co-expression occurs in diaphragm muscle fibers obviating unambiguous fiber type distinctions [16-18]. Neuregulin is a potential nerve-derived neurotropic factor that may influence contractile protein expression in muscle fibers [19, 20].

Sarcomeric Proteins

The organization of skeletal muscle from the whole muscle level down to individual sarcomeric proteins is crucial to the understanding of how muscle force is generated (Figure 2). Myofibrils are grouped into fiber bundles, that align all the sarcomeres in series to facilitate force generation. Each sarcomere forms a highly organized unit of thick (myosin) and thin (actin) filaments representing the contractile apparatus of the skeletal muscle and giving skeletal (and cardiac) muscle a striated appearance [21]. Sarcomeres are anchored and connected to neighboring sarcomeres through the Z-disc, which runs perpendicular to the filaments creating a structural and mechanical functional unit, that allows for lateral and longitudinal transmission of force and contraction. Within each sarcomere, the elemental unit of force generation is the cross-bridge, formed by the binding of myosin to actin.

Sarcomere Structure

Each sarcomere is composed of interdigitating thick and thin filaments, with dimensions of ~1 μm in diameter and ~2.5 μm in length (Figure 2). The overlap between thick and thin filaments is important in determining the number of cross-bridges, that can form during muscle activation. The thick filament has a fixed length of ~1.6 μm, while the length of thin filaments ranges between 1.0 – 1.3 μm, with the length dependent upon both the species and muscle type [22]. During muscle fiber contraction both the thick and thin filaments maintain their intrinsic lengths. However, with cross-bridge formation actin filaments are pulled toward the midline of the sarcomere, thus causing the overlap between thick and thin filaments to increase. The force-length relationship of a muscle fiber reflects the overlap between thick and thin filaments within sarcomeres and the number of bound crossbridges that contribute to force generation.

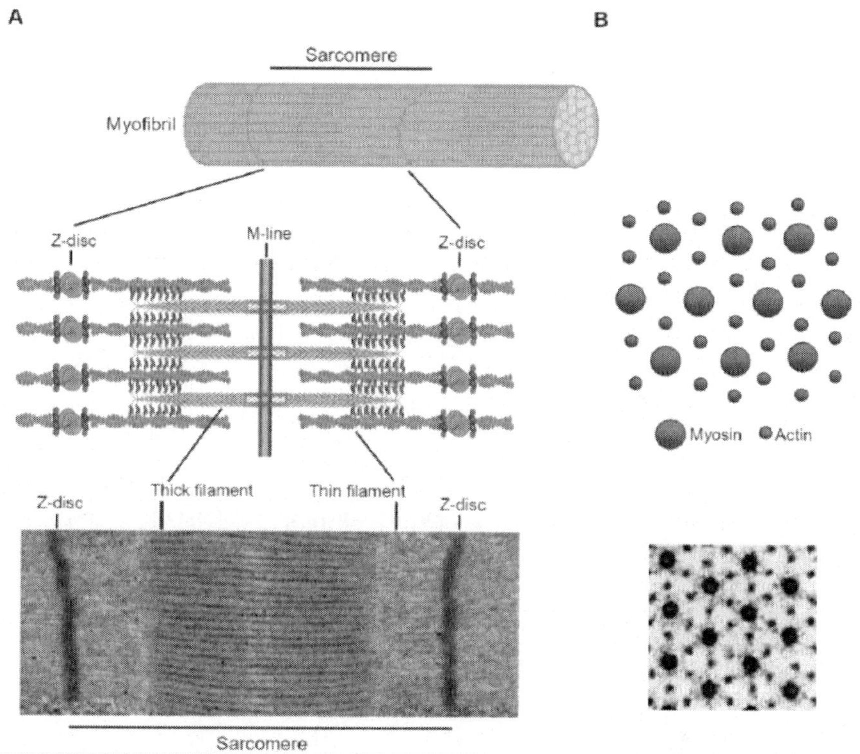

Figure 2. A) Muscle fibers comprise myofibrils that contain repeating arrangements of sarcomeres, which give muscle its striated appearance as seen by transmission electron microscopy. The primary components of the sarcomere are thick and thin filaments, which interact by crossbridge formation and slide past each other during muscle contraction. B) The crystalline organization of myosin (red) and actin (yellow) filaments creating a myofilament lattice is clearly seen in an electron micrograph of a muscle fiber cross section.

There is a fixed structural organization to the sarcomere based on the known stoichiometry between thick and thin filaments (myosin and actin proteins). A crystalline structure of thick and thin filaments is evident through electron microscopy and X-ray diffraction [23] of cross sections of muscle fibers (Figure 2). Six thin (actin) filaments surround each thick (myosin) filament, which is further surrounded by six myosin filaments with associated actin filaments. Thus, each actin filament is surrounded by three myosin filaments, and two actin filaments are shared by any given myosin filament. This arrangement creates the double hexagonal array of the myofilament lattice. The spacing between thick and thin filaments in this lattice largely determines the relative myosin and actin protein content within a muscle fiber (myosin 43-50% and actin 18-22% of total protein) [24, 25]. Since the sarcolemma (muscle fiber membrane) encompasses myofibrils in a relatively constant volume (osmolarity shifts within the muscle) due to changes in ion concentrations across the membrane, can cause osmotic compression or expansion of the lattice.Thus, it can change the distance between thick and thin filaments. Accordingly, lattice spacing is decreased (compressed) under hypertonic conditions, and increased (expanded) during hypotonic conditions. Changes in myofilament lattice spacing can affect cross-bridge attachment and thus, force generation [23].

Thick Filament Proteins

The thick filament is primarily composed of the myosin protein (specifically myosin II), which is a hexameric protein comprising two MyHCs (~220 kDa), two essential myosin light chains ($MyLC_{17}$ - 17 kDa) and two regulatory myosin light chains ($MyLC_{20}$ – 20 kDa). The arrangement of the myosin molecule is such that the tail portion is positioned toward the midline of the sarcomere and the heads are directed outward to allow binding to actin. The myosin heads are positioned in a spiral around the thick filament to align with the position of the actin filaments. The N-terminal domain of MyHC includes regions related to both the enzymatic (ATPase) and contractile (cross-bridge attachment) functions of myosin. The two MyHCs form a double helix by wrapping around each other, creating the tail region of myosin. A globular head region is located at the end of the N-terminal domain that forms the actin-binding site for cross-bridge formation. The $MyLC_{17}$ primarily plays a structural role stabilizing the myosin head; thereby enhancing cross-bridge formation and cycling. Several $MyLC_{17}$ isoforms exist, but their specific associations with different muscle fiber types are not clear at the present time.

For example, two $MyLC_{17}$ isoforms are expressed in fast-twitch (type II) fibers, but how the expression of these two isoforms relates to the mechanical properties of type II fibers is unknown. In skeletal muscle fibers, Ca^{2+}-calmodulin dependent myosin light chain kinase (MLCK) mediates phosphorylation of the serine-19 residue of $MyLC_{20}$. The effect of $MyLC_{20}$ phosphorylation in skeletal muscle fibers remains controversial, but it appears to increase cross-bridge cycling rate along with ATP consumption rate and may also affect Ca^{2+} sensitivity of force generation [26]. These effects are most likely the result of an effect of $MyLC_{20}$ phosphorylation on the position of the myosin head (moving it away from the backbone of the myosin filament) in relation to the actin filament. Several $MyLC_{20}$ isoforms also exist, but there appears to be no distinct relationship between the expression of a given $MyLC_{20}$ isoform with muscle fiber type, nor the mechanical performance of muscle fibers.

Additional proteins associated with the thick filament help to maintain the alignment of sarcomeres in parallel within and across myofibrils. The M-line is a key landmark in the midline of the thick filament, and is composed of two proteins, myomesin and M-protein, that serve to anchor titin to myosinThus, it helps stabilize and align adjacent sarcomeres. Myomesin is present in both slow-and fast-twitch muscle fibers, but M-protein is only found in fast-twitch fibers. Three isoforms of myomesin exist: one isoform (myomesin-1) exists in all fiber types, myomesin-2 is expressed only in type IIb fibers, an myomesin-3 is expressed in type IIa and IIx fibers. Titin is the largest known protein (3.2 to 3.8 MDa) and spans the sarcomere from the Z-disk to the M-line. This protein is important in stabilizing the thick filament. Titin may also function to sense the strain on the thick filament and trigger downstream signaling pathways involved in maintaining the sarcomere [27, 28]. Several isoforms of titin exist due to alternative splicing [29].

Generally longer titin isoforms are expressed in slow fibers, and it has been suggested that this might impart differences in passive stiffness across muscle fiber types, that could affect contractile performance. Myosin binding protein C is also located at regular intervals along the length of the thick filament protein. Ithas binding sites for the myosin rod and titin. It is thought to be involved in assembling the backbone of the thick filament. Two isoforms of myosin binding protein C exist in skeletal muscle, associated with slow and fast fibers.

Thin Filament Proteins

The major contractile protein of the thin filament is α-actin, a globular protein (40 kDa) that contains the binding sites for myosin to form cross-bridges. Although different isoforms of α-actin exist, their relationship to fiber type or mechanical performance remains unclear. Globular actin monomers (G-actin) assemble into helical double stranded chain of filamentous actin (F-actin), which forms two strands of an alpha helix. Between six and seven globular actin monomers are attached to tropomyosin (TM), which in turn is associated with a troponin complex, that together regulate exposure of the myosin binding site on actin. The thin filament also includes the nebulin protein, which is a large protein (0.6 to 0.9 MDa) that spans the length of the thin filament and is thought to regulate actin filament length [30]. It has been suggested that variants in both nebulin and titin may affect passive and active mechanical properties of muscle fibers.

Excitation of a muscle fiber is initiated by a nerve action potential and is transmitted through the neuromuscular junction. Once an action potential is initiated in a muscle fiber, it is propagated along the sarcolemma and eventually depolarization is passively transmitted down deep transverse invaginations of the plasma membrane known as transverse tubules (t- tubule). This depolarization is coupled to Ca^{2+} release from the sarcoplasmic reticulum via activation of dihydropyridine receptor (DHPR) channels at the triad junction. The release of Ca^{2+} from the sarcoplasmic reticulum triggers cross-bridge formation through an interaction with the troponin/TM complex on the thin filament.

Neuromuscular Junction

The neuromuscular junction (NMJ) is the synapse between a motor neuron and the group of muscle fibers it innervates (Figure 3), which together constitute a motor unit. The NMJ comprises pre- and postsynaptic elements. At the presynaptic terminal of a motor neuron, synaptic vesicles contain acetylcholine (ACh) as a neurotransmitter to effect neuromuscular transmission.

Synaptic vesicles dock with the presynaptic terminal membrane at specialized regions called active zones. The number of docked vesicles at each active zone is comparable across fiber types [31].

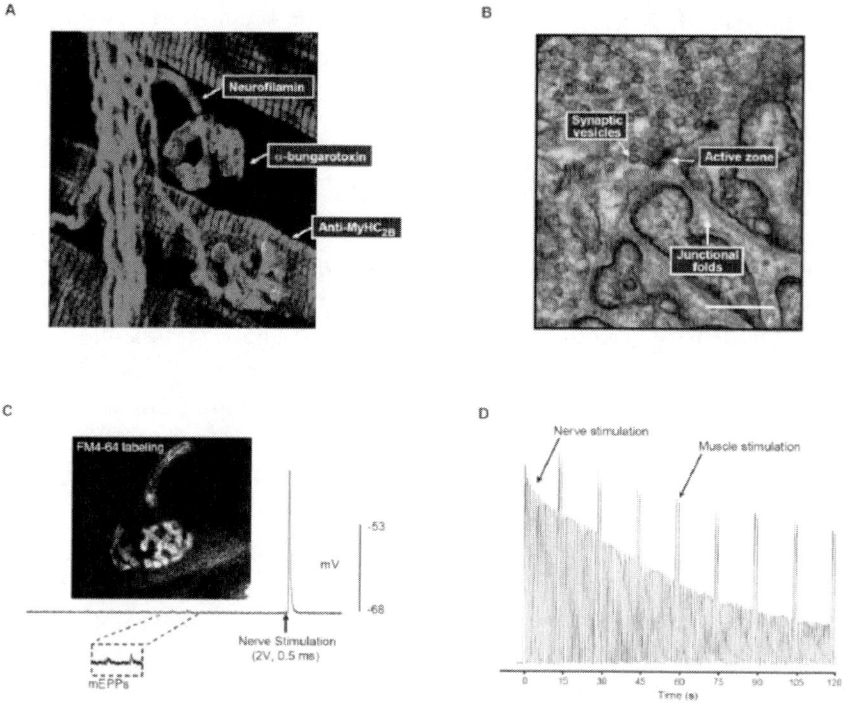

Figure 3. A) Representative confocal image of neuromuscular junction (NMJ) showing motor axons down to the presynaptic terminal (immunoreactivity for neurofilamin), motor endplates (cholinergic receptors fluorescently labeled using α-bungarotoxin) and muscle fibers (labeled with an antibody specific to the MyHC2B isoform). B) Electron micrograph showing the preand postsynaptic elements of a NMJ. Adapted from 31, used with permission. C) Electrophysiological recording of spontaneous miniature end-plate potentials (mEPPs) and an evoked EPP response. The presynaptic terminal was visualized by uptake of a styryl dye FM4-64. D) Neuromuscular transmission failure induced by repetitive nerve stimulation is reflected by the difference in muscle force generated by nerve stimulation compared to that induced by periodic (every 15 s) direct muscle stimulation.

However, the total surface area of the presynaptic terminal is smaller at type I and IIa fibers compared to type IIx and IIb fibers. Thus, the total number of active zones at presynaptic terminals is greater at type IIx and IIb fibers, and accordingly, the total number of synaptic vesicles released in response to a nerve action potential (quantal content) is greater at these fibers compared to type I and IIa fibers [31]. The motor end-plate is a specialized region of postsynaptic muscle fiber membrane that contains nicotinic cholinergic receptors, that are activated by presynaptic release of ACh. To increase the surface area, there is both branching and postsynaptic folding at the motor endplates. This postsynaptic branching and folding is far more complex at type IIx and IIb fibers compared to type I and IIa fibers [32]. The motor end-plates at type I and IIa fibers are further distinguished by the imposition of mitochondria, rough endoplasmic reticulum, and free polysomes between the postsynaptic specializations and myofibrils [32]. There are no apparent differences in the density of postsynaptic cholinergic receptors across fiber types; however, given the larger surface area of the postsynaptic membrane at type IIx and IIb fibers, the total number of cholinergic receptors is greater at these fibers. Accordingly, endplate potential responses to a nerve action potential-induced release of ACh is greater at type IIx and IIb fibers. In addition, the density of voltage-gated Na^+ channels adjacent to the motor end-plate is higher at type IIx and IIb fibers, than at type I and IIa fibers [33]

Thus, the membrane depolarization required to initiate a muscle fiber action potential (threshold) is lower at type IIx and IIb fibers [34]. Together, the greater end-plate potential response and lower threshold account for a higher safety factor for neuromuscular transmission at type IIx and IIb fibers. However, with repetitive stimulation, the amplitude of the evoked end-plate potential response decreases markedly in type IIx and IIb fibers,and with it the safety factor for neuromuscular transmission [34]. Accordingly, these fibers are much more susceptible to neuromuscular transmission failure during repetitive stimulations compared to type I and IIa fibers [35].

T-Tubule and Triad Junction

An action potential generated at the NMJ is propagated down the muscle fiber membrane (sarcolemma) by the activation of voltage-gated Na^+ channels. There does not appear to be any fiber type differences in the propagation of muscle fiber action potentials. The t-tubule membrane contains a lower density of voltage-gated Na^+ channels; thus, depolarization is passively conducted down the t-tubules, where it activates DHPR channels, which are specialized voltage-sensing Ca^{2+} channels. In skeletal muscle fibers, it is unlikely that significant Ca^{2+} influx occurs when DHPR channels are activated. Instead, the DHPR channels allosterically interact with ryanodine receptor (RyR) channels located in the terminal cisternae of the sarcoplasmic reticulum that directly about the t-tubule at the triad junction. During depolarization of the t-tubule, DHPR channels undergo a conformational change, that is mechanically linked to the opening of RyR channels at the triad junction, allowing Ca^{2+} release from the sarcoplasmic reticulum. There is no evidence that there are fiber type differences in excitation-contraction coupling at the level of the t-tubule and triad junction.

Sarcoplasmic Reticulum and the T System

To have a basis for comparison, it will be advantageous to review briefly the evolution of present concept of the relationship of the sarcolemma to the internal membrane systems of the sarcoplasm in skeletal muscle [481]. Each myofibril of skeletal muscle was shown by Bennett and Porter [36] in 1953 to be surrounded by a lacelike network of tubular elements comprising the sarcoplasmic reticulum. The predominant orientation of the component sarcotubules was longitudinal; in works on amphibian muscle, it was found that these tubules were confluent with conspicuous transverse elements of larger caliber called terminal cisternae [37], situated on either side of the Z line. Between the parallel terminal cisternae was a row of small tubular and vesicular profiles. The apparent discontinuity of these membrane-limited intermediate elements later proved to be artifactitious [38] and, with improved fixation, a slender continuous tubule -40 mμ in diameter could regularly be demonstrated between the pairs of terminal cisternae [39]. The lumen of this tubule was subsequently shown to be in open communication with the extracellular space [40-47]. Individually the slender transverse elements, which are continuous with the sarcolemma are called T *tubules* and collectively they constitute the T *system* of skeletal muscle. Although they were originally considered to be a part of the sarcoplasmic reticulum, they are now regarded a separate system of invaginations of the sarcolemma.

A transverse tubule of the sarcolemma and the two associated terminal cisternae of the sarcoplasmic reticulum comprise the socalled *triad* of skeletal muscle (Figure 4).

In fish and amphibians, the triads are at the level of the Z lines, but in mammals they occur, two to a sarcomere, at or near the A-I junctions. It is now widely accepted that the T tubules of the triads provide for spread of electrical activity from the surface of the muscle fiber into its interior [48, 49, 50], while the terminal cisternae and the contiguous longitudinal sarcotubules of the reticulum are probably involved in release and recapture of calcium ions in the contraction-relaxation cycle [51, 52, 53].

Early investigators of the fine structure of the heart found none of the triads characteristic of skeletal muscle, but recognized, in thin sections, pairs of membrane-limited structures consisting of a narrow elongate profile closely apposed to the periphery of a larger circular opening.

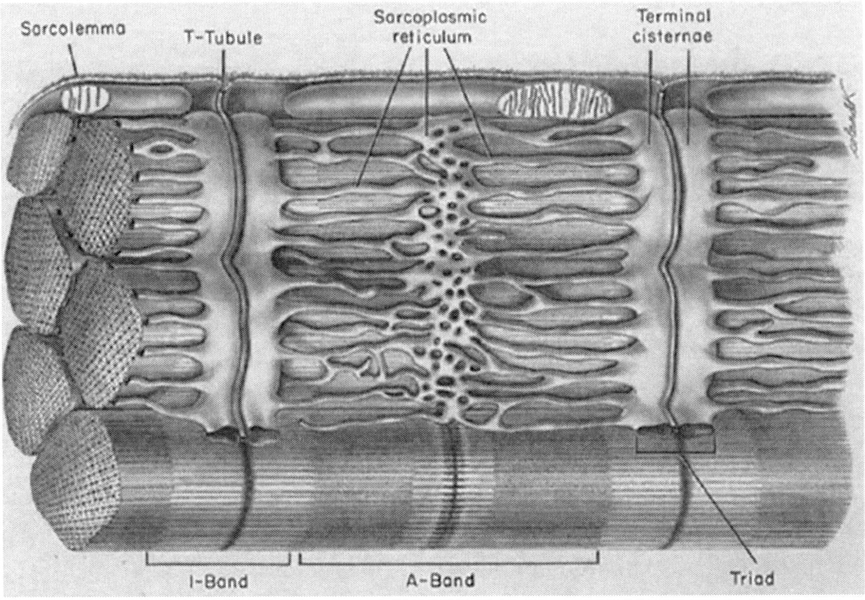

Figure 4. A drawing depicting the T tubules and sarcoplasmic reticulum relation to several myofibrils of amphibian skeletal muscle. There are discrete myofibrils of uniform size, each ensheathed by sarcoplasmic reticulum, with pairs of terminal cisternae associated at each Z line with a slender T tubule to form a "triad ". The localization of the triads at the Z line of amphibian muscle makes it more comparable to mammalian cardiac muscle [481].

These structures were termed "dyads", in the belief that they corresponded to the triads of cardiac muscle [54]. Some investigators erroneously interpreted the larger profile, as corresponding to a single terminal cisterna and the smaller, to the intermediate element of the striated muscle triad [55].

Continuity between a system of transverse tubular structures and the extracellular space of striated muscle was actually first suggested by Lindner [56] on the basis of electron microscopic studies of cardiac muscle, but this observation was largely overlooked. Subsequent investigations of heart muscle of several species have now clearly established, that it is the larger element of the so called dyads of cardiac muscle that corresponds to the intermediate tubule of the triads of skeletal muscle. The continuity of these relatively large transverse tubules with the cell surface has been repeatedly demonstrated [57-60].

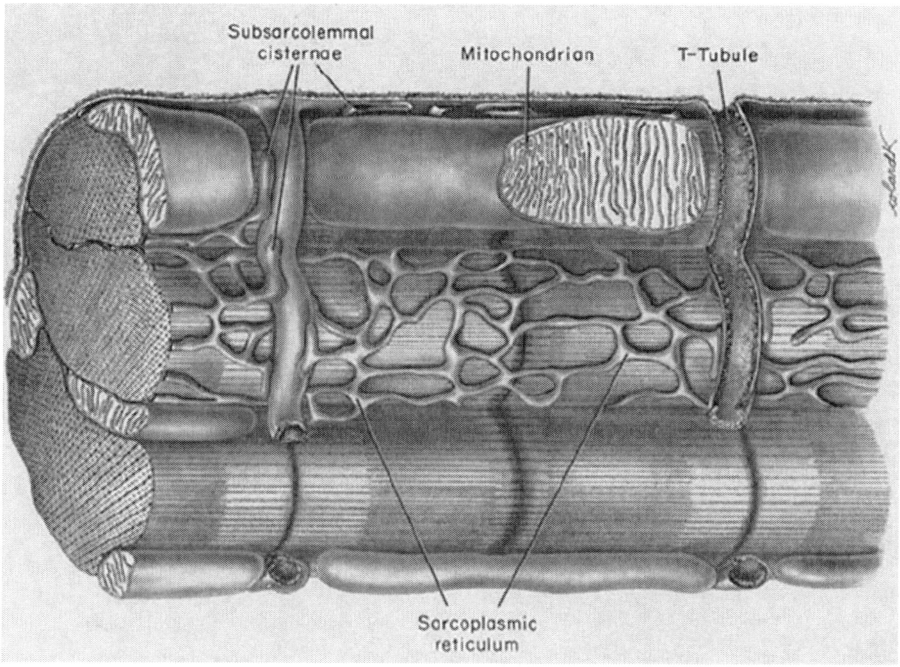

Figure 5. A drawing of the T tubules and sarcoplasmic reticulum of mammalian cardiac muscle. Notice the large size of the T tubules, the simpler pattern of the reticulum, and the absence of terminal cisternae. Instead of terminal cisternae, small saccular expansions of the reticulum, called subsarcolemmal cisternae, are in close contact with the T tubes or with the sarcolemma at the periphery of the fiber. The myofilament mass is partially subdivided into irregular myofibril-like areas by clefts that are penetrated by T tubules. (Drawing previously published [481].

The diameter of T tubules of cardiac muscle varies somewhat with the method of specimen preparation, but it is of the order of 1500-2000 A as compared to -400 A for the T tubules of skeletal muscle.

Owing to the straighter course and larger size of T tubules, their openings at the cell surface are encountered in electron micrographs of cardiac muscle more frequently than in electron micrographs of skeletal muscle. Occasionally, the openings are found at several successive Z lines in the same section (Figure 6 and 7), and the inference can be made that they tend to occur in rows. This is borne out in recent observations by the freeze-etch method, which show that the apertures of the transverse tubules appear on the cell surface in approximately parallel rows (89). These rows are interpreted as corresponding longitudinally to the clefts in the myofilaments occupied by the sarcoplasmic reticulum, and transversely, to the in-register Z-band regions.

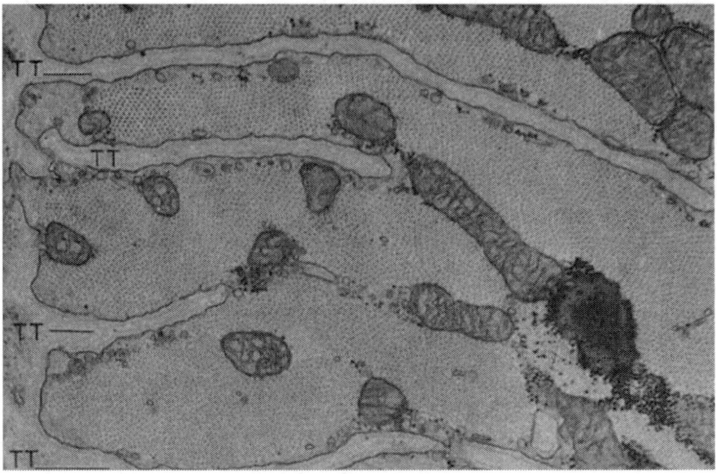

Figure 6. Micrograph of a transverse section of cardiac muscle showing four T tubules (TT) extending inward from the periphery of the fiber. This is an uncommon picture for the reason that the plane of a thin section seldom happens to coincide with one of the rows of T tubes opening onto the surface. The upper three T tubules are somewhat unusual in being located at the level of the A band, but the lower one is at the level of the I band as expected. X 32,000.

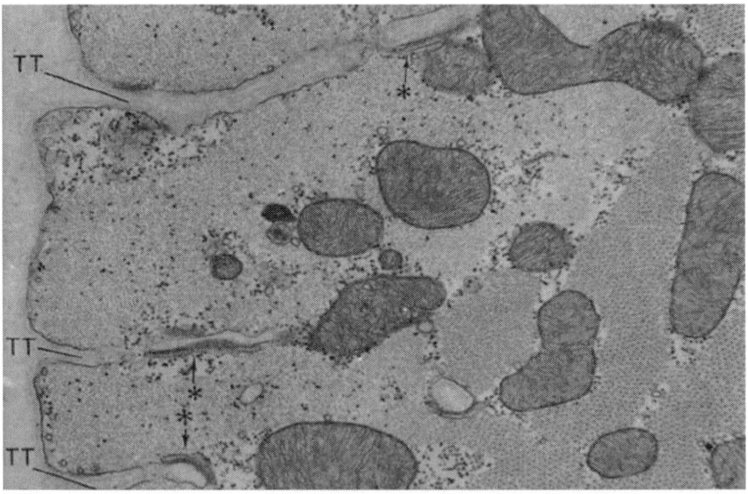

Figure 7. A transverse section at the periphery of a cardiac muscle fiber showing the origin of three T tubules at the level of theI band. This section also coincides with one of the rows of T tubule openings (TT) at the periphery of the fiber. Both the sarcolemma and its protein polysaccharide coat are invaginated to form the T system which is clearly a direct extension of the extracellular space into the "interior" of the cell. Three subsarcolemmal cisterns associated with T tubules are indicated by arrows and asterisks X 35,000.

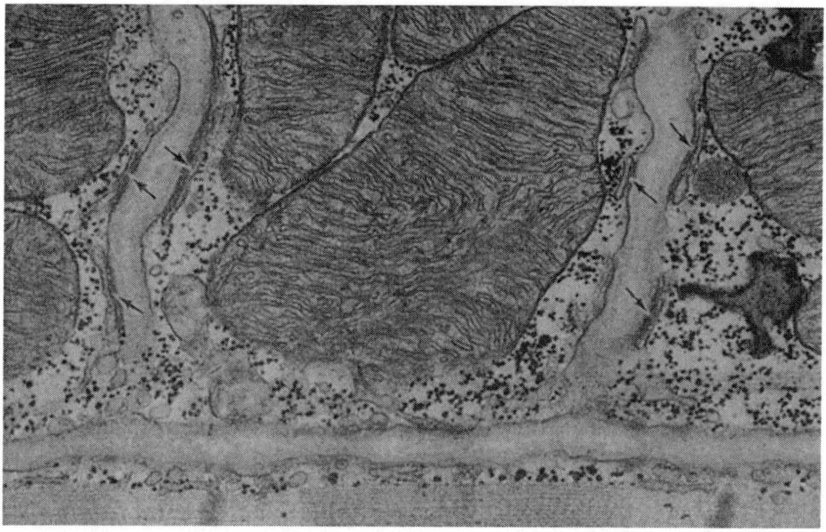

Figure 8. Longitudinal section of papillary muscle showing two T tubules and at the arrows, several sites of contact with subsarcolemmal cisternae of the reticulum. The T tubules curve out of the plane of section, and their openings onto the surface are not shown here, but are in the neighboring section. X 40,000.

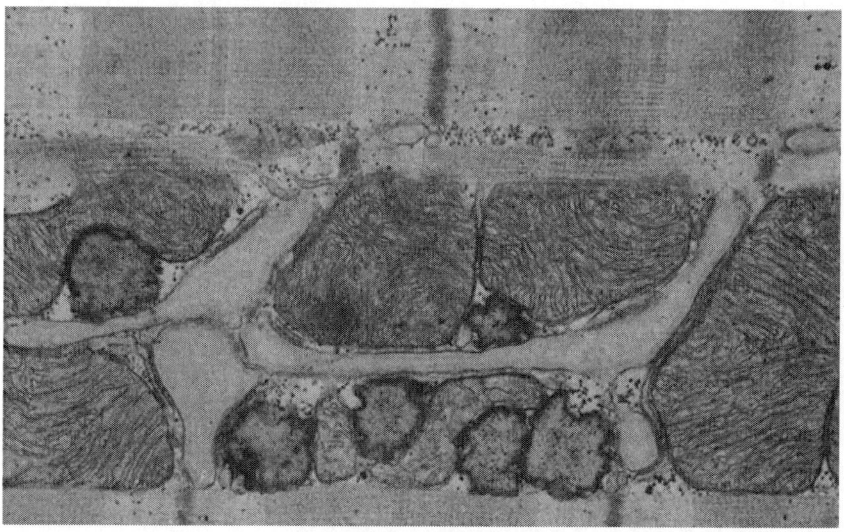

Figure 9. Longitudinal section of the central region of a muscle cell illustrating two longitudinal interconnections between successive T tubules. Although as the name implies, T tubules generally run transversely, longitudinal branches such as are shown here occur quite frequently. X 33,000.

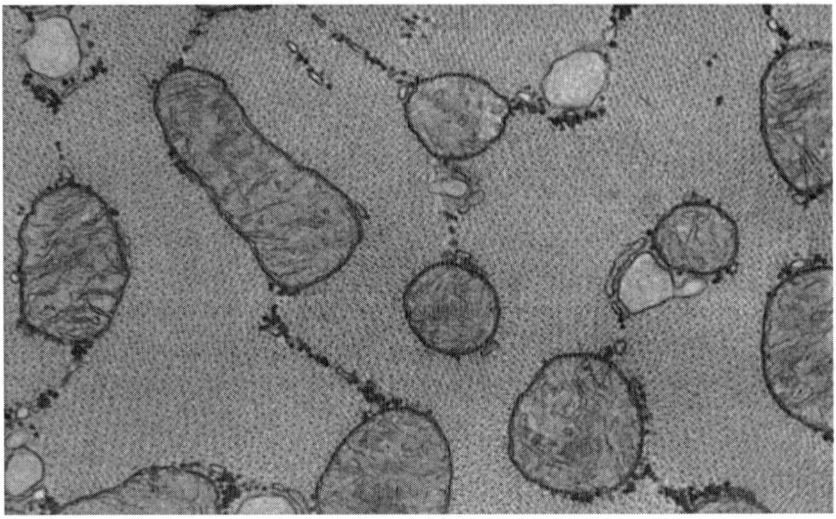

Figure 10. Transverse section of papillary muscle fiber showing five rounded profiles of longitudinal branches of T tubules in the relatively small area encompassed by this Figure. Two of the branches have subsarcolemmal cisternae associated with them. Longitudinally oriented branches of the T tubules are, therefore, more common than is generally realized. X 54,000.

At the mouths of the transverse tubules, the protein-polysaccharide coating of the sarcolemma can be traced without interruption into the tubules (Figure 6), and its presence is helpful in identifying profiles of this system, deep in the interior of the fiber (Figure 10 and 11).

As noted by Simpson and Oertelis [61], additional evidence that the sarcolemma and the membrane lining the T system are identical is to be found in the frequent occurrence, on both structures, of small vesicular invaginations of the kind, commonly interpreted as a form of micropinocytosis. The depth of penetration of the T tubules into the cell is, no doubt, variable, but it is not uncommon to find sections of them near the nuclear envelope, or in the axial core of sarcoplasm. Thus, the length of T tubules may be as much as half the diameter of the fiber. The term "transverse tubule" accurately describes their prevailing direction, but it is somewhat misleading in that it tends to obscure the fact that there are also branches of this system that run longitudinally and connect successive T tubules (Figure 8).

In longitudinal sections, these longitudinal T tubules are not easily distinguished from extracellular clefts between adjacent muscle cells, but in transverse sections of heart muscle their circular profiles are easily identified and they are found to be quite numerous (Figure 9).

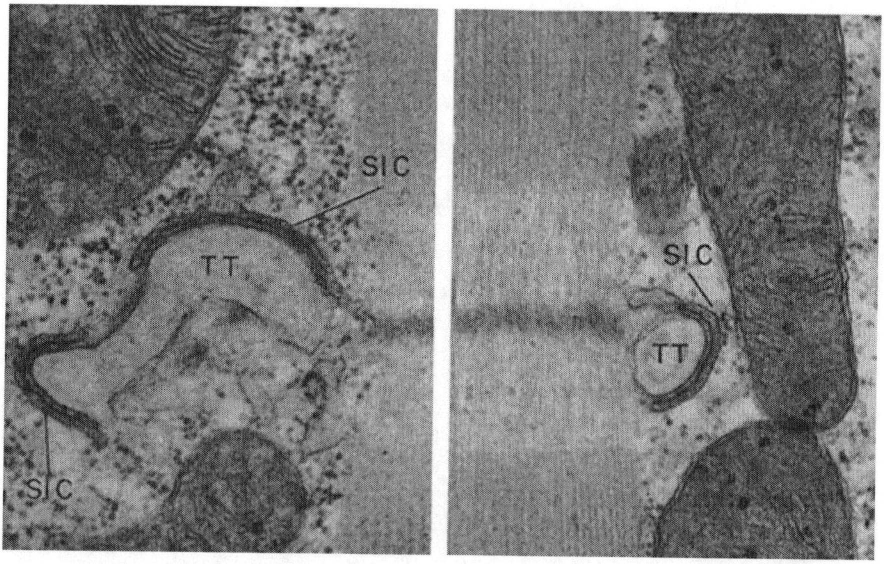

Figures 11-12. Longitudinal sections of papillary muscle showing typical examples of flattened saccules or cisternae of the reticulum (SIC) closely applied to T tubules (TT). With glutaraldehyde primary fixation the cisternae are rather flattened and appear to have a dense content. X 57,000.

Longitudinal connections between the transverse tubules of different sarcomeres were described for skeletal muscle by Veratti [62] and Revel [63]. Their occurrence in cardiac muscle was noted by Nelson and Benson [64], and Simpson and Rayns [65], but they have not received the attention that they deserve. The sarcoplasmic reticulum of cardiac muscle is less extensive than that of skeletal muscle, and its plexiform pattern is not locally modified in relation to particular bands of the repeating cross-striations (Figure 4). The pairs of transverse cisternae located at the Z bands, or at the A-I junctions of skeletal muscle are absent. There is no region of especially abundant lateral anastomosis of the sarcotubules at the middle of the A band (compare Figure 4 and 5). Instead, the sarcoplasmic reticulum of cardiac muscle is a simple plexus of tubules of rather uniform caliber, freely anastomosing at all levels of the crossbanded pattern (Figures 39 and 40).

In the absence of discrete myofibrils, it ramifies over the dering course of these clefts, extensive surface of the large coherent bundle of myofila- views of the reticulum are uncommon in cardiac ments and extends deep into narrow clefts. (Figures 14-16).

A study of such areas reveals that there are no transverse cisternae and no other interruptions in the longitudinal continuity of the sarcotubular network either at the Z line or at the A-I junctions. At the I-band region, however, flattened saccular expansions of the sarcotubules are closely applied to the sarcoplasmic surface of the T-tubule (Figsure 4, 7, 9-11).

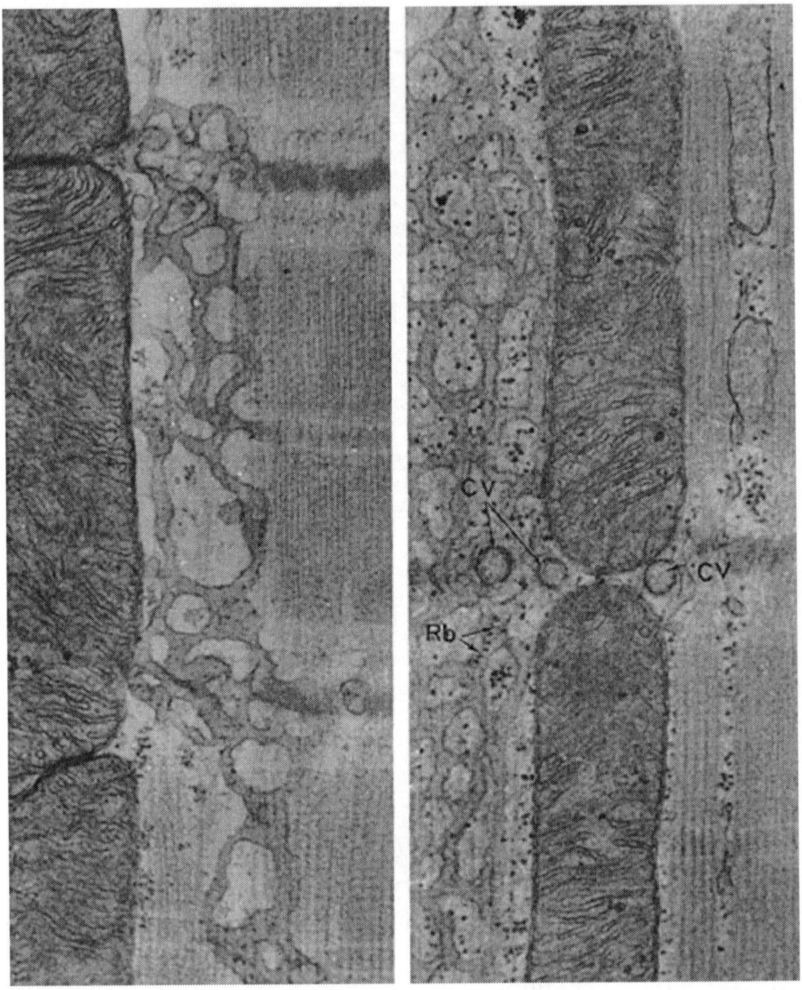

Figures 13-14. Two views of the sarcoplasmic reticulum which consists of a simple network of anastomosing tubules, continuous across the Z line and with no local specialization there or over the center of the A band. The circumferential "Z tubule" described by some authors as encircling myofibrils at the level of the Z disc is not a regular occurrence in the cat. In Figure 14 are three examples of the coated vesicles (CV) that are very often located at the Z line. At the upper part of the Figure (arrows and asterisks) the membrane limiting a portion of the reticulum has a similar density and surface coat, suggesting that one or more coated vesicles have become incorporated into the wall of the reticulum. X 47,000: X 51,000.

These flattened *saccules*, or "foot plates" of some authors [66-68], are the smaller element in the "dyads" of earlier descriptions of cardiac muscle. They are also applied to the inner aspect of the sarcolemma at the periphery of the fibers and occasionally to small areas of the surface of contact between successive muscle cells. Although of much more limited content, these local expansions of the sarcotubules are the anatomical, and presumably the physiological, counterpart of the terminal cisternae of skeletal muscle. Their relationship to the sarcolemma is identical, whether they occur at the periphery of the fiber or at the T tubules in its interior. The term subsarcolemmal cistern is perhaps the most useful term for these structures, for it is equally descriptive of those associated with the T system and those at the periphery of the fiber. The flattened saccule or cistern is parallel to the sarcolemma for a distance of from 0.1 to 1.0 μm and is separated from it by a 150-200 A space which has a beaded or serrated appearance owing to the presence of globular densities -150 A in diameter, regularly spaced about 200 A apart (Figure 17).

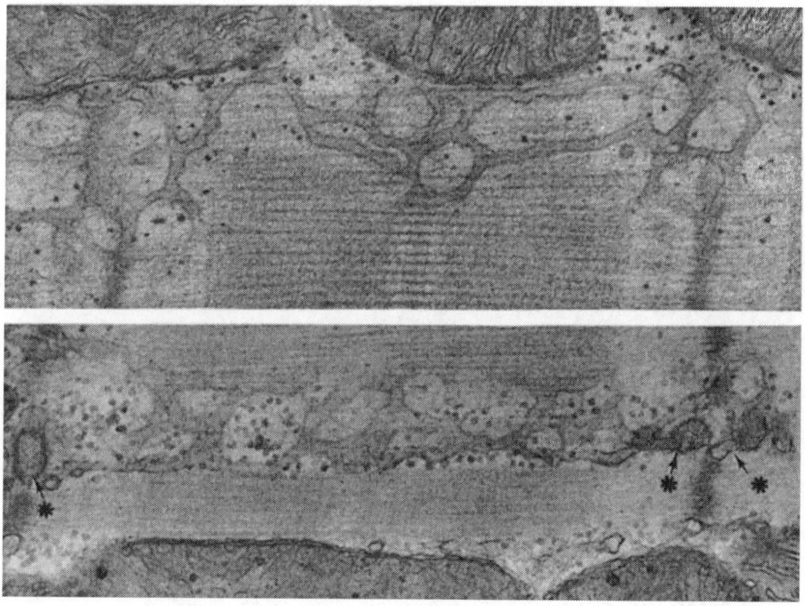

Figures 15-16. Two additional examples illustrating the continuity of the reticulum across the Z line, with no specialized circumferential element that could be designated a Z tubule. Figure 16 shows, at the asterisks and arrows, examples of coated vesicles with dense content associated with the reticulum at the Z line. The two vesicles at the right appear to be continuous with tubular elements of the reticulum. X 50,000 ; X 56,000.

These ill-defined densities between the sarcolemma and the adjacent surface of the reticulum cannot be said to be more closely associated with one membrane than the other. The subsarcolemmal cistern has a content which often appears as a 30-50 A dense line in the center of its 150 A flattened lumen (Figure 11 and 17). The sarcoplasmic reticulum of cardiac muscle is very largely composed of smooth-surfaced membranes and is considered to be a special form of the agranular endoplasmic reticulum. However, short segments of sarcotubules may bear polyribosomes on their surface [69] (Figure 13). These ribosome-studded tubules appear as rather randomly distributed segments in the reticulum. If the membranes are continually being renewed, as appears to be the case in the smooth reticulum of other cell types [70], then these rough-surfaced segments intercalated in the system of smooth sarcotubules may play an important role in the biogenesis of the new smooth membrane.

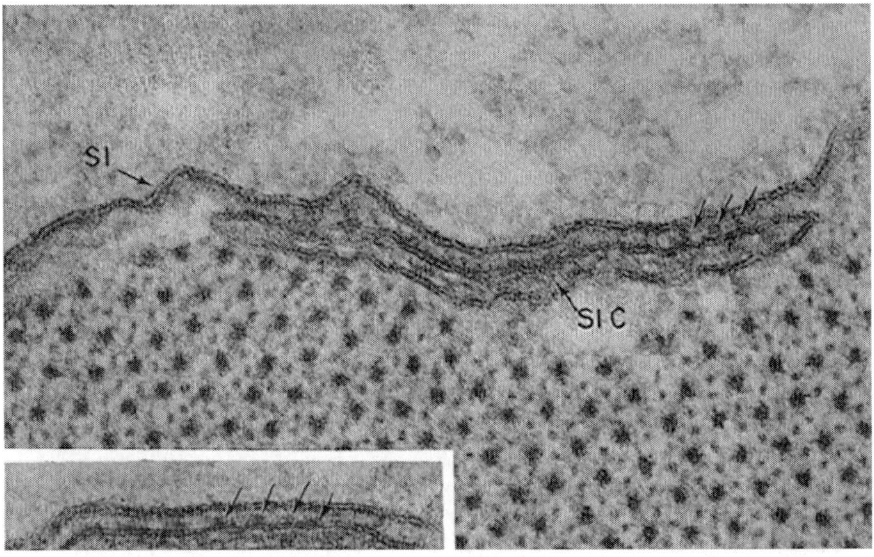

Figure 17. High-magnification micrograph of the sarcolemmal unit membrane (SI) and a subsarcolemmal cisterna (SIC). The limiting membrane of the latter is separated from the sarcolemma by a distance of 150-200 A. Periodic densities in this interspace (arrows) give it a beaded or serrated appearance. The content of the saccule or cisterna is often condensed to form a central linear density. The inset presents another example showing to better advantage the periodic densities (arrows) between the subsarcolemmal cistern and the sarcolemma. X 300,000.

These segments are probably more prevalent than is commonly appreciated, since glycogen particles of similar size occurring in abundance in the same areas tend to interfere with their visual recognition in electron micrographs. In cardiac muscle of the ox [70] and the ferret [67], a slender circumferential tubule 220-400 A in diameter has been described as closely encircling the "myofibrils" at the level of the Z disc.

This structure has been called the Z tubule and is said to communicate with the longitudinal sarcotubules of the reticulum and laterally with the Z tubules of other "myofibrils." Although both occur at the level of the Z band, the Z-tubule is said to be easily distinguishable from the T tubules because of its smaller size. According to Simpson and Rayns [67], the Z tubules in the species they have studied form the primary connection of the sarcoplasmic reticulum to the flattened saccules that are in apposition to the T tubules. In the optimally stretched cat myocardium, Z tubules could not be clearly identified. In longitudinal sections providing *en face* views of the reticulum, its meshes are seen to continue uninterruptedly from one sarcomere to the next (Figure 12-16). No continuous tubule has been found running parallel to the Z line such as depicted by Simpson and Rayns [67]. Moreover, in transverse sections, slender circumferentially oriented tubules were not observed at the level of the Z band more frequently than at other levels. Therefore, in the cat myocardium there are no regularly occurring circumferential tubules at the Z bands and hence no category of sarcotubules that deserves the separate designation Z tubule. Previously undescribed components commonly found in the vicinity of the Z bands are spherical, coated vesicles 400-600 A in diameter (Figure 13 and 16). In the character of their limiting membranes, these vesicles resemble the coated vesicles found in the Golgi region and at the cell surface. They usually appear free in the sarcoplasm, but not infrequently they connect by narrow necks to tubular elements of the sarcoplasmic reticulum near the Z lines (Figure 16). Occasionally, short segments of sarcotubules of the reticulum near the Z band have a limiting membrane that displays a similar finely fibrillary coating (Figure 13). The frequent localization of coated vesicles at or near the Z bands remains unexplained. The occasional continuity of these vesicles with the sarcotubules in this region suggests, however, that they either arise from or coalesce with the reticulum. Their very common occurrence at the Z band suggests the possibility that these coated vesicles may have contributed to the impression of other authors that circumferential tubules (Z tubules) are common at this site. Since they are a striking finding in the atrial myocardium, these vesicles are described in more detail in the companion paper [71].

Comparative Ultrastructure of Ca^{2+} Release Units in Skeletal and Cardiac Muscle

Understanding the topology and geometry of muscle membranes and the spatial relationship between the key molecules is essential to unraveling the mechanism of regulated Ca^{2+} release during activation of muscle contraction [482]. Important clues are provided by functional and structural similarities as well as differences between skeletal and cardiac muscle. SR and exterior membranes interact with each other at intracellular junctions, named triads, dyads and peripheral couplings on the basis of their geometry (Figure 18).

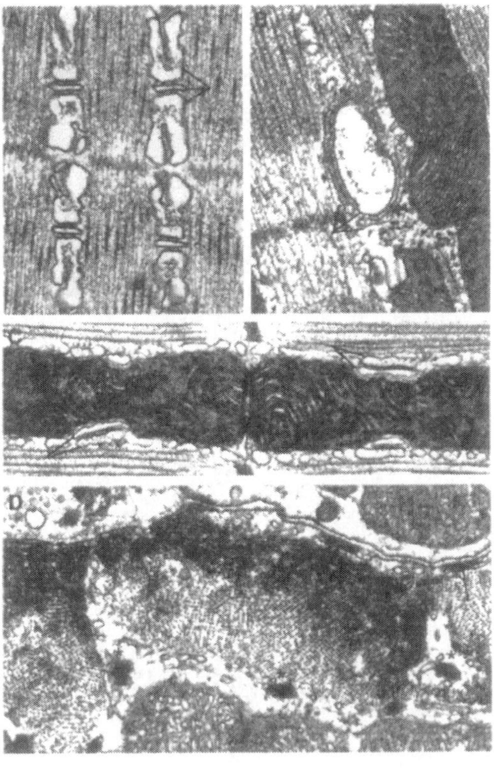

Figure 18. Examples of junctions between the sarcoplasmic reticulum and exterior membranes with a variety of geometries. Triads (three lines) are junctions between two SR elements and one segment of the T tubule network. A. Four triads in a skeletal muscle fiber from the toadfish swim- bladder. The transverse tubules are flat ribbons with two narrow

surfaces facing wide SR cis- ternae. The junctional surfaces are oriented transversely to the direction of the filaments. B. A triad in a cardiac muscle celi from the mouse left ventricle. The T tubule has a large diameter, and two flat SR cisternae surround it. The junctional surfaces are oriented longitudinally, or par- allel to the direction of the filaments. C. Dyads (two lines) in the flight muscle of the dragonfly, located between the mitochondria and the myofibrils. The junctional surfaccs are oriented lon- gitudinally. D. Peripheral couplings (single lines), or junctions between the SR and the surface membrane in a cardiac celi from the chicken left ventricle. Other SR elements in the fiber inte- rior (asterisks) have junctional components (feet and calsequestrin) but are not associated with exterior membranes. These are called corbular ancl/or extended junctional SR. Magnification: 50,000x.

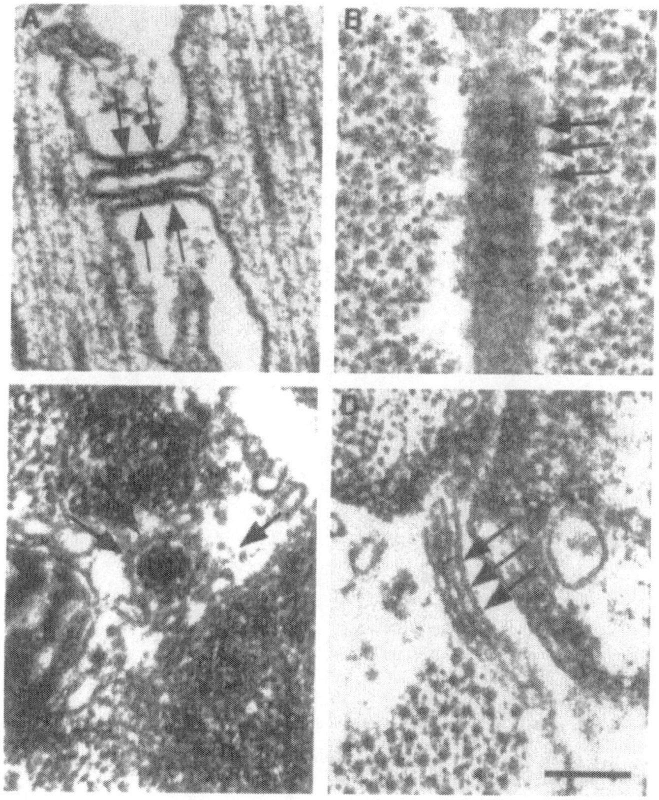

Figure 19. Ali SR elements participating in junctions (A,B, and D), and those forming the cor- bular SR (C) have feet (ryanodine receptors) associated with the junctional membrane. A and B: sections across and tangent to triads in the toadfish swimbladder. Arrows point to feet forming two rows within the junctional gap separating SR and T tubuie membranes. C and D: corbular SR and a peripheral coupling from chick myocardium, showing periodically disposed feet (arrows). The dense content of the SR is calsequestrin. Magnification: 133,000x, bar = 0.1 μm.

The Ca^{2+} release channels of the SR, also called ryanodine receptors, are located at the junctional sites and are visible as feet in the electron microscope (Figure 19A, B, D).

In cardiac muscle, a special SR arrangement, containing the feet, but not forming junctions with the surface membrane (the corbular or extended junctional SR, EjSR), is also present (Figure 19C). The T tubules and surface membrane participating in the junctions contain clusters of L-type Ca^{2+} channels or dihydropyridine receptors (Figure 20A and B).

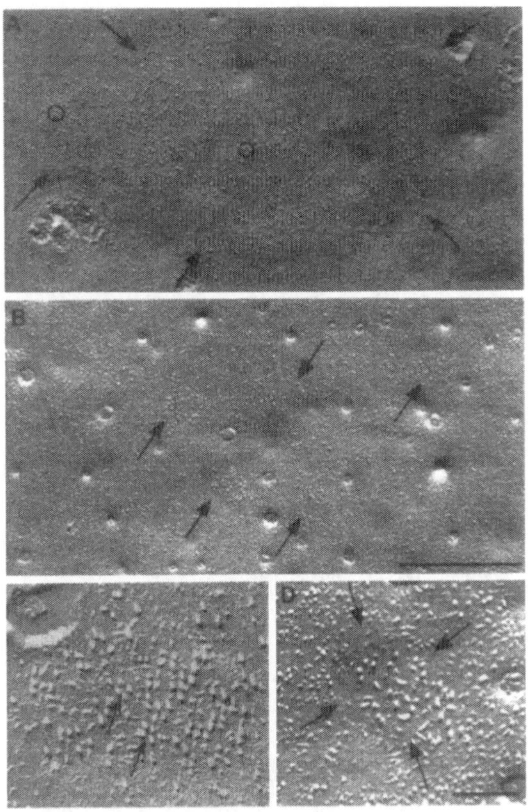

Figure 20. Dihydropyridine receptors (DHPRs) are detectable as tall intramembranous parti- cles of large diameter in freeze-fracture replicas. Clusters of DHPRs are illustrated on the surface membrane of BC3HI, a cell line of skeletal muscle origin (A, between arrows); on the surface of chick ventricular myocardium (8, at arrows; D between arrows); and developing mouse skeletal muscle in vivo (C). DHPRs are clustered in both skeletal and cardiac muscle, but in a difTerent manner. In skeletal muscle, DHPRs form arrays of tetrads (C, arrows point to individuai tetrads), while in cardiac muscle DHPRs are randomly disposed. Magnification: A,B: 51,000x, bar = 0.5 μm; C,D: 135,000x, bar = 0.1 μm.

RyRs and DHPRs are located in correspondence to each other in discrete junctional domains of SR and exterior membranes, forming functional units named e-c coupling units or Ca^{2+} release units [72, 73]. The junctional domains constituting Ca^{2+} release units are in turn separated from each other by nonjunctional domains of the two membranes, which contain a different complement of proteins. Thus the units are arranged in focal spots along the continuous exterior membranes. Modeling indicates that this discrete structural arrangement may result in the concerted action of severa Ca^{2+} release channels within functional units, called couplons [74]. The Ca^{2+} release units in skeletal and cardiac muscle and the muscles of invertebrates have different geometries. In adult skeletal muscle of most vertebrates, transversely arranged T tubule networks prevail and the usual junction is a triad, with a transverse orientation and two or at most three elongated rows of feet (Figure 19A and B). During skeletal muscle differentiation, in the period between the initial formation of myofibrils and T tubule development, peripheral couplings are the only type of junction present. As soon as T tubules develop, however, longitudinally and, later, transversely arranged triads and dyads appear, and peripheral couplings become rare. A few fiber types, such as frog slow tonic fibers, maintain a high frequency of peripheral couplings. In muscles of arthropods, the junctions are mostly triad or dyads with a longitudinal orientation (Figure 18C), and feet are arranged in a round plaque. Peripheral couplings may also be frequent. In cardiac muscle, T tubules have a much larger diameter than in skeletal muscle, and the most common junctions are in the form of dyads formed by the dose apposition of a flat SR cisterna, which contacts the T tubule over a wide area. Occasionally a triad is present (Figure 18B), but it has a different appearance from the skeletal muscle triad, due to the wider lumen of the T tubule. Peripheral couplings (Figure 18D and 19D) are often present in parallel with internal junctions. All myocardial cells in avian and frog hearts and in the atria of other vertebrates, lack T tubules and have fre- quent peripheral couplings. In addition, avian muscles are particularly rich in extended junctional SR, the avian equivalent of the mammalian corbular SR (Figure 18D, 19C). These variations in the geometry of junctions affect the location, distribution, and overall quantity of the Ca^{2+} release channels within the muscle fiber. However, the specific mechanism by which Ca^{2+} release is initiated, it is not likely to differ between triads, dyads, and peripheral couplings within the same muscle type. EjSR, on the other hand, must be considered separately.

All Ca^{2+} release units have feet disposed in highly ordered arrangements, with very similar spacings. Minor differences in the orientation of feet have been detected between vertebrate and arthropod muscles. Furthermore, it is not known whether the disposition is exactly the same in cardiac and skeletal muscles. These minor differences may not affect the function of the channels. The disposition of DHPRs, on the other hand, is significantly different in cardiac muscle and muscles of invertebrates on the one hand, and in skeletal muscle on the other. In skeletal muscles, DHPRs are grouped into tetrads, or clusters of four DHPRs, located at the corners of small squares. The tetrads in turn are disposed in larger ordered arrays, so that the four components of the tetrads are located in correspondence of the four subunits of the feet [75,76]. This location is present only in skeletal muscles, and it constitutes the structural basis for a unique direct molecular interaction, that allows the voltage sensor of e-c coupling (the DHPR) [77, 78] to directly affect the permeability of the SR Ca^{2+} release channel (the RyR or foot), in the so-called mechanical model of e-c coupling [79]. The four components of the tetrads may act in concert [80].

Neither cardiac DHPR nor RyR can substitute for their skeletal isoforms in establishing a direct molecular interaction with their skeletal counterpart [81, 82]. Direct confirmation that DHPRs and RyRs are functionally coupled in skeletal muscle come from the demonstration, that a retrograde signal from RyRs is necessary for full functioning of DHPRs [83]. In cardiac muscles, DHPRs are also clustered in dose proximity to RyRs [84], but they do not form tetrads and are not disposed in a detectable ordered arrangement and therefore are not likely to establish a link to the RyRs [85, 86]. DHPR-RyR interaction in these muscles is dependent on extracellular Ca^{2+} and on Ca^{2+} current through the DHPRs, and so it is thought to be indirectly mediated by Ca^{2+}. The dose DHPR-RyR proximity in cardiac muscle is appropriate, if RyR opening is regulated by the Ca^{2+} current flowing through the DHPRs, perhaps through Ca^{2+} ions. The disposition of DHPRs in muscles of arthropods is similar to that of cardiac muscle, [87] and this also is in keeping with the fact that e-c coupling in these muscles may be mediated by Ca^{2+}-activated Ca^{2+} release [88, 89]. An additional difference between cardiac and skeletal muscle is that while in the former all assemblies of feet are associated with DHPR-bearing membranes, in the latter the feet can also be independently located in clusters (within corbular and EjSR), which are located deep within the muscle fiber, at relatively large distances from either the surface of the cell or the T tubules [90, 91]. These internal Ca^{2+} release units have the same RyR isoform as the peripheral couplings [92]. Thus, while in skeletal muscle a command from DHPRs may be the only means of initiating a Ca^{2+} release response from a cluster of feet, in cardiac muscle, a Ca^{2+} release unit may be activated without intervention of DHPRs.

In cardiac muscles that lack T tubules, activation of the internal Ca^{2+} release units is necessarily dependent on diffusion of a messenger, presumably Ca^{2+}, from one unit to the other. Modeling of this interaction is critically dependent on knowing the distance between one unit and its immediate neighbor [93]. In avian myocardium, we compared the minimum distance, separating peripheral couplings at the fiber surface from each other, to the minimum distance between EjSR elements, located within a narrow transverse layer at the level of the Z line. The minimum distance between adjacent peripheral couplings, 472 ± 5 nm (mean ± I SEM, n = 340), is significantly larger than the average distance for EjSR, 135 ±15 nm (n = 263). Interestingly, the minimum distance between adjacent Ca^{2+} release units (triads) along the T tubules of three representative skeletal muscle fibers-fast twitch fibers from frog semitendinosus and ileofibularis and from male toadfish swimbladder muscle-are quite small (IO I ± 2 nm, n = 225 for both frog muscle and 107 ± 2 nm, n = 358 for the toadfish). A comparative look at skeletal muscle shows that the distances between Ca^{2+} release units vary depending on fiber types within the same animal. The units in fast twitch fibers are separated by very short stretches of nonjunctional T tubules (see above), while in slow twitch and even more so in slow tonic fibers, the distances can be very large [94-96]. One of the direct results of large spacings between Ca^{2+} release units is of course the fact that the total concentration of feet is smaller, and thus c-e coupling results in a smaller Ca^{2+} release. In the fast muscle of the toadfish swimbladder, where Ca^{2+} release units are very close to each other and thus the overall density of feet in the fiber is large (Figure 21). Consequently, the Ca^{2+} transients are very fast and very large [97]

In skeletal muscle fibers of different types, Ca^{2+} release units are not only spaced at variable intervals along the T tubules, but also have variable sizes. In general, the faster the muscle, the larger the c-e coupling unit, and thus the larger the content of feet and DHPRs. The average number of feet per unit (or, in this case, triad) varies between 112 for the superfast toadfish swimbladder muscle, and 20 for the frog slow tonic fiber. Units in guinea pig fast twitch oxidative, fast twitch oxidative-glycolytic, and slow twitch fibers average 68, 48, and 34 feet each, and the frog twitch fibers of somewhat mixed character have 72 (iloeofibularis) and 58 (sernitendinosus) feet/triad. One notable exception in the trend for larger units in faster muscles, is the superfast male toadfish swimbladder muscle, in which triads have 36 feet on the average, a number sirnilar to that in the much slower tonic fibers of the frog. In cardiac muscle, the size of Ca^{2+} release units may also vary, from 29-30 feet per unit in chick myocardium, to 150-200 feet/unit in mouse and rat left ventricle.

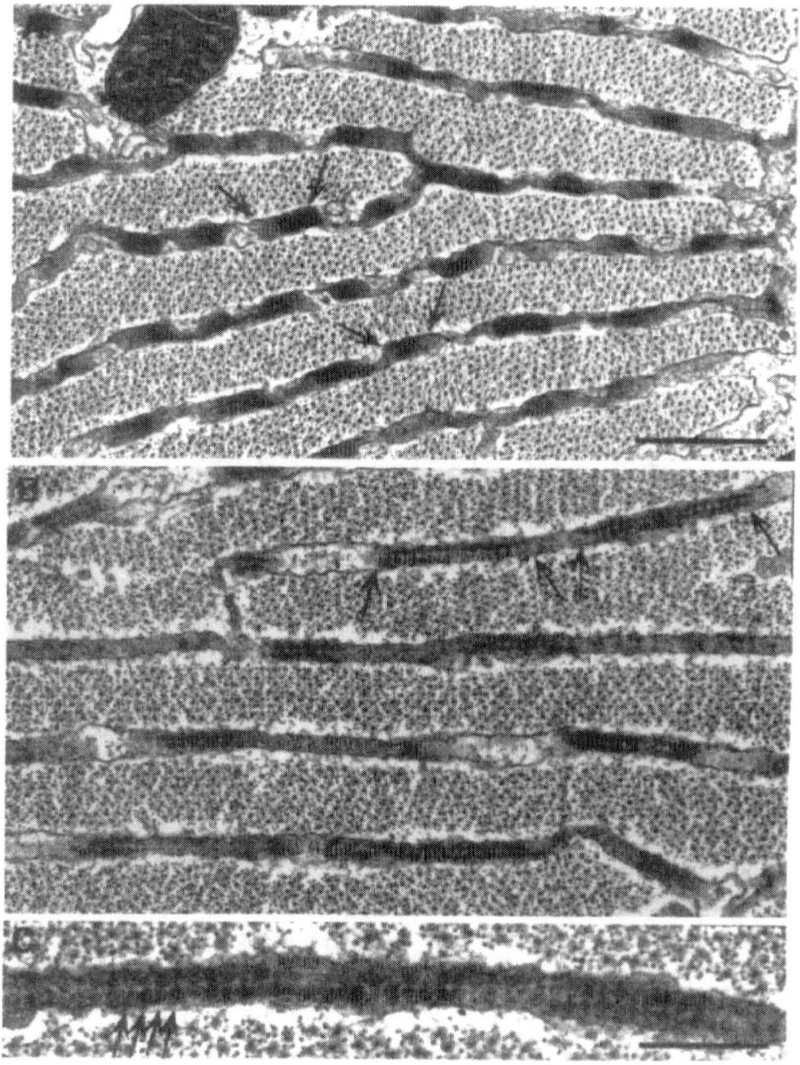

Figure 21. The size of calcium release units varies greatly in skeletal muscles, depending on the length of the T tubule segment that is engaged in association with the SR. In general, the size of the release units is related to the fiber type: fast muscle fibers have larger calcium release units (see text). The superfast toadfish swimbladder muscle, illustrated here, is an exception, because the muscle in the male swimbladder has short junctions (A, between arrows), while the junctions in the same muscle from the female has longer junctions (B, between arrows; C, arrows point to couples of feet). In the male, short junctions bear at most 24 feet on either side of the T tubuie, while in the female an individual junction may contain as many as 115 feet on one side. Magnification: A,B: 35,000x, bar = 0.5 μm; e, 70,000x, bar = 0.25 μm.

Thus some Ca^{2+} release units in cardiac muscle have a larger content of feet than the units in very fast skeletal muscles, despite the slower rate of activity of the muscle fibers. One possible explanation for this unexpected observation may be that the apparently single, large units in cardiac muscle may in reality be an aggregatc of smaller, closely spaced units.

Structure and Location of Sarcoplasmic Reticulum in Smooth Muscle

Although it may sound outmoded to be concerned with the structure and location of the SR, recent work on membrane microdomains and SR compartmentalization create a clear factual understanding of SR structure and distribution in order to better judge what is feasible, both within the SR and in its relation to other organelles and the plasma membrane. Much of our structural information concerning smooth muscle has come from EM studies. These studies addressed the size of the SR, usually compared with striated muscle and in terms of cell volume, and SR location within myocytes. By revealing these features EM made a significant contribution not just to the understanding of SR structure, but also to E-C coupling in smooth muscle. Before fluorescent probes sensitive to ion concentration were developed, electron X-ray microprobe analysis provided spatial information about the elemental composition of different areas in the myocytes, in relaxed and contracted stages. Consistent between different workers and smooth muscles is the location of the SR: close to, but not touching the plasma membrane. The SR was noted to encircle the invaginations of the surface membrane, the caveolae. Sheets of reticulum running along the longitudinal axis of the cell are a feature of all the tissues examined, e.g., vascular [98], airway [99], gastrointestinal tract and bladder [100], human blood vessels [101], developing blood vessels [102], and portal vein [103]. The SR is described as peripheral (or superficial) when it is close to the plasma membrane and central (or deep) when it is located away from this membrane. The peripheral SR has been associated with Ca^{2+} homeostasis, local Ca^{2+} release, and interactions with plasma membrane ion channels and hence excitability, whereas the central SR has been suggested to be more directly involved in contraction by supplying Ca^{2+} to the myofilaments. Given the different activity and mechanisms of excitation between phasic and tonic smooth muscles, differences in the amount and localization of the SR have been sought. In vas deferens, a phasic muscle, the SR at EM level was described as being located predominantly at the periphery, whereas the tonic aorta had proportionally more SR in a central location [104]. The smooth muscle cell's interior is packed with myofilaments and intermediate filaments. It is estimated that they, along with associated dense bodies, occupy up to 90% of the cell's volume. The intracellular organelles share the remaining volume [105]. Although the earliest EM accounts described the SR of smooth muscles as being poorly developed [106, 107, 108], this view gradually changed and is borne out by recent studies using confocal microscopy. In one of the best comparative studies, Devine et al. [98] compared rabbit smooth muscles and reported that 5% of the cell volume was accounted for by the SR in aorta and pulmonary artery and that this fell to 2% in mesenteric vein and artery, and portal vein. These findings also correlated with the varying ability of these preparations to continue contracting in zero Ca solutions, i.e., longer in aorta than mesenteric and portal vein. However, when low values for SR volume were reported in some (*vas deferens* and *taenia coli*, 1.7%) but not all (*uterus*, 7%) [109] phasic smooth muscles, the correlation with contraction time in zero Ca fell away. Somlyo's group pointed out that osmium fixation can damage smooth muscle SR and that this could have contributed to reports of being sparse [110]. Thus the view gradually changed from "However, we must bear in mind that a sarcoplasmic reticulum, as we understand the structure from skeletal muscle work, is definitely not present in smooth muscles, and it seems most unlikely that the minute fraction of reticulum present in some smooth muscles is capable of modulating calcium levels" [111] to "Our major conclusion is that there is in mammalian smooth muscles a sarcoplasmic reticulum that, while being variable in different smooth muscles, is sufficiently well developed and organized to function as a Ca^{2+} store in the process of excitation-contraction and inhibition-relaxation coupling" [98]. Notwithstanding the fact that smooth muscle SR is well developed, its volume is less than that of striated muscles (10% of cell volume). Therefore, there are reasonable grounds to question whether the SR in smooth muscle can have the same role as in striated muscles, particularly with respect to it being the amplifier of a small Ca^{2+} entry signal to provide the necessary rise of $[Ca^{2+}]$ required for contractile activation by CICR. This questioning is also suggested by the generally smaller SR volume in visceral smooth muscles, where Ca entry is predominant, compared with vascular smooth muscles, where SR Ca release is more prominent. It seems to us that the *pendulum swung* from initially dismissing the SR's role in modulating Ca signals and CICR in smooth muscle to viewing the SR as playing a very similar role to that occurring in striated muscles. We are perhaps now able to readjust the *pendulum* and see more clearly that smooth muscle SR is important to E-C coupling, but in its own distinctive way. In summary, recent data obtained by live staining of native smooth muscle cells have added confidence to earlier data on SR structure and distribution obtained with EM. There is no certainty about what the distinction between peripheral and central SR means, in terms of SR components or functional consequences for any particular smooth muscle. Perhaps because of the considerable interest in local Ca signals from the SR and their relation to plasma membrane ion channels, the literature tends to emphasize the peripheral SR, and few studies have addressed the role of central SR. All phasic smooth muscles examined with confocal techniques appear to have predominantly peripheral SR. As well as the interactions between the SR and plasma membrane and caveolae, there is also interest in elucidating the functional importance for smooth muscles of having their SR enveloping mitochondria and appearing con-tinuous with the nuclear envelope.

Chapter 2

Development of the Sarcoplasmic Reticulum

Abstract

During excitation-contraction (e-c) coupling of striated muscle, depolarization of the surface membrane is converted into Ca^{2+} release from internal stores. This process occurs at intracellular junctions characterized by a specialized composition and structural organization of membrane proteins. The coordinated arrangement of the two key junctional components-the dihydropyridine receptor (DHPR) in the surface membrane and the ryanodine receptor (RyR) in the sarcoplasmic reticulum is essential for their normal, tissue-specific function in e-c coupling. The mechanisms involved in the formation of the junctions and a potential participation of DHPRs and RyRs in this process have been subject of intensive studies over the past years [483].

Vesicular Transport Systems in Differentiating Skeletal Muscle Cells

Skeletal muscle cells are multinuclear cells with a very long cylindrical appearance, with the longest human skeletal muscle cells having a length of up to 14 cm [112]. The smallest human skeletal muscle cells are found in the stapedius muscle and their length is 2 mm, which is 100-fold longer than the average mononucleated cell (10–20 μm) [113]. Skeletal muscle cells have contraction ability and are equipped with a specialized plasma membrane and intracellular structures, such as neuromuscular junctions, myotendinous or myo–myo junctions, myofibrils, T-system, and sarcoplasmic reticulum SR. These specialized structures of skeletal muscle cells develop in a multiple-step process, that consists of (1) the proliferation of mononuclear myoblasts, (2) the fusion of myoblasts to form multinuclear myotubes, and (3) the maturation of myotubes. Systematic vesicular transport is expected to play a role in the differentiation of skeletal muscle cells from myoblasts through to myotubes and for the maintenance of the mature skeletal muscle cells [114]. However, most of the research on vesicular transport in differentiating cells has been performed on neurons and epithelial cells, and few studies have focused on skeletal muscle cells.

In the early 1960s satellite cells were identified as resident myoblasts in adult muscular tissues [115], and subsequent ultrastructural studies in the 1960s and 1970s described the morphological characteristics of these satellite cells. Satellite cells in adult animals are less active mitotically and metabolically than those in juvenile animals based on their characteristics of heterochromatic nuclei, poorly developed Golgi *apparatus*, and low levels of glycogen [116]. The satellite cells of developing muscle [117] and young animals [116] are more active, based on the presence of euchromatic nuclei and a well-developed rough endoplasmic reticulum ER. Satellite cells have caveolae and pinocytotic vesicles in their cytoplasm [117, 118]. The satellite cells are considered to be myogenic stem cells, and activated satellite cells, referred to as myoblasts, initiate the muscle differentiation process. Research on the differentiation of skeletal muscle cells shifted from morphological studies to molecular biology studies in the 1990s, with the latter revealing a number of transcription factors [119] and adhesion molecules [120], which regulate the differentiation of skeletal muscle cells. However, to date, no serious effort has been made to determine the intracellular features of myotubes.

The myotubes, multinucleated muscle fibers that form from the fusion of myoblasts, are longer than typical mononuclear cells, usually growing to a length of 0.1–0.6 mm even in culture [121]. In a recent electron microscopy study, we examined the organelles in whole myotubes as a basis for studying membrane traffic in myotubes [122]. We found that the perinuclear region of the myotubes contained small-sized Golgi *apparatuses* and, that autolysosomes and mitochondria were abundant in the perinuclear region. The rough ER extended from the perinuclear region toward the periphery of the myotube, while the smooth ER was located further from the nuclei. A large number of vesicles were located near the tip of the myotubes. Vesicles were observed in both satellite cells (myoblasts) and myotubes. These vesicles are expected to play an important role in the intracellular vesicular transport system, for delivery of membrane components and including proteins to the appropriate organelles.

To identify the nature of these vesicles in the vesicular transport system, it could be extremely helpful to be focused on soluble N-ethylmaleimide-sensitive factor attachment protein receptor (SNARE) proteins and describe their protein expression profiles in myoblasts and myotubes. Expression profiles of the VAMPs during muscle differentiation are summarized in Table 1.

Table 1. Expression of vesicle-associated membrane proteins during muscle differentiation [483]

VAMPs[a]	Intracellular localizations in non-muscle cells	VAMP expression during muscle differentiation			
		Satellite cell (quiescent)	Myoblasts (proliferating)	Myotubes	Skeletal muscle cells (matured)
VAMP1	SV	–	–	–	–
VAMP2	SV, etc.	+	–	+	+
VAMP3	EE, RE	–	+	+	+
VAMP4	TGN	NR	+	+	NR
VAMP5	PM	–	+	+	+
VAMP7	LE, Ly, PM	NR	+	–	–
VAMP8	EE, LE	NR	+	+	NR

SV Synaptic vesicles, EE early endosome, RE recycling endosomes, TGN trans-Golgi network, PM plasma membrane, LE late endosome, Ly lysosome, NR not reported.

[a]VAMPs, Family of vesicle-associated membrane proteins belonging to the category of v-SNAREs (N-ethylmaleimide-sensitive factor attachment protein receptors).

Proliferating myoblasts (C2C12 mouse myoblast and/or primary culture myoblasts) express VAMP3, 4, and 8. VAMP3 and 8 may play roles in endosomal trafficking, and VAMP4 may be involved in TGN trafficking. Myotubes express VAMP2, 3, 4, 5, and 8 (Figure 22).

VAMP2, 3, and 5 may regulate endosomal trafficking at the periphery of the myotubes, with VAMP2 and VAMP3 showing similar localization and VAMP5 localized solely near the plasma membrane. These observations may reflect the multiple steps of endosomal trafficking between the cytoplasm and plasma membrane. VAMP4 in myotubes is involved in TGN trafficking at the perinuclear region, while VAMP8 in myotubes shows diffuse signals and therefore may not play significant roles in myotubes. These data were obtained by static morphological analyses and, therefore, are not supported by information on time-dependent changes in the localization of VAMPs, or the molecular interactions of VAMPs with other molecules. VAMPs interact with t-SNAREs to form the SNARE complex and function to tether and fuse vesicles to the target membrane [123]. The formation of the SNARE complex is also regulated by other molecules, such as small GTPases [124]. Only a few studies have reported the expression and the distribution of small GTPases in differentiating skeletal muscle cells. ADP ribosylation factor (ARF) 1 regulates vesicle budding from the ER to the SR [125]. ARF6 is distributed at the plasma membrane and influences myoblast fusion [126, 127]. Rab8A and Rab13 are involved in insulin-dependent GLUT4 translocation in skeletal muscle cells [128]. However, the interactions between VAMPs and other molecules have not been identified in myoblasts or myotubes. Future studies on the molecular interactions and molecular dynamics of VAMPs will clarify the route and target of the vesicles in myotubes.

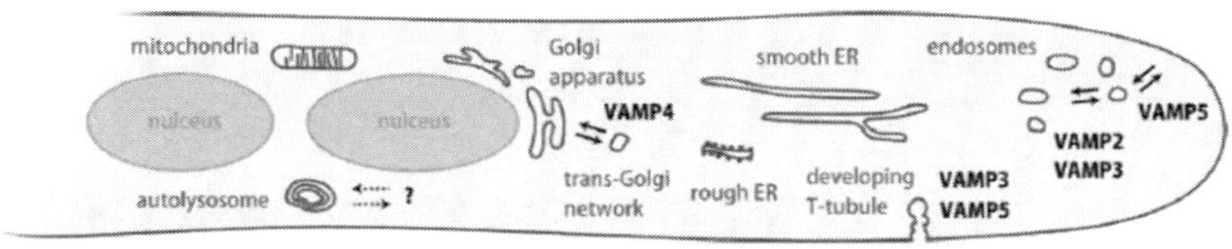

Figure 22. Organelles and VAMPs in myotubes [483]. To show the organelles and VAMPs clearly, we omitted differentiating myofibrils in this Figure. Mitochondria and autolysosomes are abundant around the center of the cell. The Golgi *apparatus* and trans-Golgi network labeled with VAMP4 are localized in the perinuclear region. ER expands from the perinuclear region toward the periphery. VAMP2, 3, and 5 are localized at the periphery of the myotubes, with VAMP2 and 3 localized similarly, and VAMP5 localized closer to the sarcolemma. Portions of VAMP3 and 5 are localized at the developing T-tubules.

Further studies are required to address how the VAMP bearing vesicles contribute to the differentiation of skeletal muscle cells. VAMP3 and VAMP5 are colocalized with caveolin-3, a marker of developing T-tubules, leading us to suggest that VAMP3 and VAMP5 may be involved in T-tubule development [122]. In addition to T-tubules, characteristic structures, such as neuromuscular junctions [129] and myotendinous junctions [130], should be investigated in future studies, although many difficulties are encountered when attempting to reproduce a whole differentiation process of the skeletal muscle in experimental model systems.

Myoblasts in culture are available as cell lines (C2C12, BC3H1, and MM14 from mouse and L6 from rat) or can be isolated from rodent and avian muscle tissues [131]. These cells do not differentiate to show the full contractile function and the generation of different fiber types. Animal models can be used to study muscle differentiation and do show the terminal differentiation, but it is hard to analyze the time-dependent changes of a single skeletal muscle cell. Research on vesicular transport in myotubes has just marked its beginning. Further expansion of this field will yield insight into not only the nature of skeletal muscle cells but also muscular diseases.

Formation of Junctions Involved in Excitation-Contraction Coupling in Skeletal and Cardiac Muscle

The recent analysis of the molecular constituents of triads and their organization in developing normal and diseased muscle has led to a new understanding of the developmental sequence and the molecular mechanisms involved in triad formation. Unexpected was the finding that interactions between the chief molecular constituents of the triad, DHPR and the ryanodine receptor RyR, do not play a primary role in the initial association of the SR with surface membranes (plasma membrane and T tubules), although they may be involved in the tissue-specific assembly of the DHPR within the junctions. Thus, during junction formation an initial membrane docking step can be dissociated from mechanisms involved in the specific targeting and immobilization of the receptors in the junction [484].

The signal transduction process occurring at the triad involves: (1) depolarization of the plasma membrane and its invaginations, the T tubules [132]; (2) a charge movement across the membrane [133] corresponding to the voltage sensing of the dihydropyridine-sensitive L-type calcium channel DHPR [134, 135]; and (3) release of Ca^{2+} from the SR stores through the ryanodine-sensitive Ca^{2+} release channel RyR [136, 137]. In both skeletal and cardiac muscle, activation of the SR Ca^{2+} release channel is under the tight control of the DHPRs, resulting in a rapid and efficient coupling of membrane depolarization and SR Ca^{2+} release [132, 138].

However, the mechanism of e-c coupling differs in the two muscle types. In skeletal muscle, gating of the Ca^{2+} current through DHPRs is slow and apparently not required for triggering SR Ca^{2+} release, whereas in cardiac muscle activation of the L-current is fast and necessary for e-c coupling [139]. Thus it has been proposed that the main function of the DHPR in skeletal muscle is that of a voltage sensor, which may control the opening of the RyR through direct molecular interactions [133]. In cardiac muscle, on the other hand, the Ca^{2+} current through the DHPR seems to mediate the interaction between DHPRs and RyRs by a process called Ca^{2+}-induced Ca^{2+} release [140-142].

The structural organization of DHPRs and RyRs in the junctions of skeletal and cardiac muscle reflects both similarities and differences in the e-c coupling mechanisms of the two muscles (Figure 23).

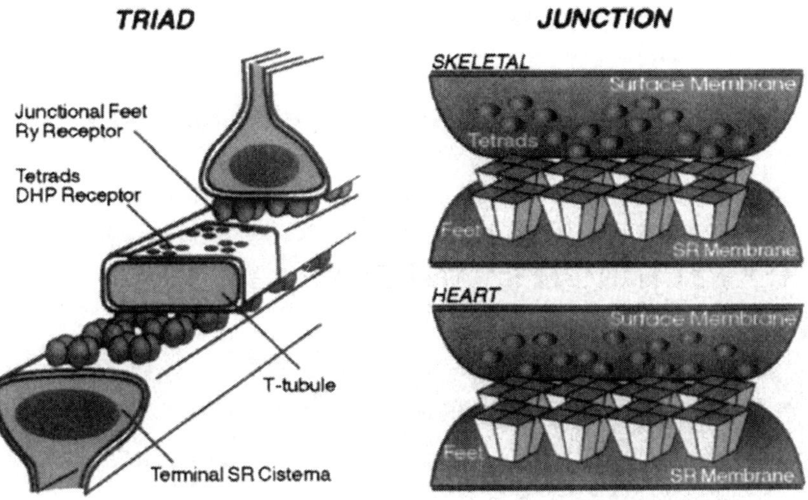

Figure 23. Structure of junctions between SR and T tubule/plasma membrane [484]. A schematic view of the triad (Left) shows the position of the cytoplasmic domain of RyRs (feet) in the junctional gap between the T tubule and the two terminal SR cisternae and the position of DHPRs (tetrads) in the protoplasmic face of the junctional T tubule membrane. A representation of the junctional gap-as if the adjacent membranes were peeled apart-shows the matching disposition of feet and tetrads in skeletal junctions (Right, Upper) and the irregularly clustered DHPR particles opposite the feet in cardiac junctions (Right, Lower). Note that tetrads in skeletal junctions correspond in size and orientation to the four subunits of the feet, but that only every other foot is associated with a tetrad.

The most prominent structural feature of all junctions are the "feet," which are electron-dense structures periodically spanning the narrow junctional gap that separates the membranes of SR and either T tubules or the plasma membrane [143]. In rotary shadowed preparations of isolated SR vesicles, an individual foot appears as four equal spheres forming a semicrystalline array in the plane of the membrane [144]. The four spheres correspond to the cytoplasmic domains of the hometrameric RyRs [137]. High-resolution image reconstruction of the RyR reveals the four-fold symmetry of the channel-forming transmembrane domains and of the cytoplasmic domains [145-147]. The latter has a complex system of pores and cavities, which presumably serve to direct the efflux of Ca^{2+} into the junctional gap.

Despite sequence differences, skeletal and cardiac RyRs share the same structural characteristics and a position in the junctional gap, which allows the close interaction with components of the junctional surface membrane [148]. Junctional domains of T tubules and the plasma membrane contain a single population of large intramembrane particles, revealed by freeze-fracture studies. The disposition of the particles, however, differs significantly in skeletal and cardiac muscle.

In skeletal muscle, the particles are clustered in groups of four, called tetrads [149]. The arrangement of tetrads corresponds closely to that of the feet. This indicates that the two classes of molecules may actually be linked to each other, allowing each of the components of a tetrad to interact with one of the four subunits of the feet (Figure 23). The localization of the tetrads in junctional domains of T tubules and the plasma membrane coincides with the localization of high concentrations of DHPRs [150, 151]. Identification of the tetrads with the DHPR is further supported by the lack of both in the dysgenic mouse mutant [152] and their concurrent restoration by transfection of the mutant cells with cDNA encoding the skeletal muscle DHPR a, subunit [153].

The close spatial relationship of tetrads and feet is a prerequisite for a direct functional interaction of DHPRs and RyRs in skeletal muscle e-c coupling. Surprisingly, however, the spacing between tetrads is twice that of the feet, resulting in two populations of RyRs-those that are associated with the DHPR and those that are not associated with DHPR. It is not certain whether both categories of feet act as Ca^{2+} release channels. In cardiac muscle, DHPRs are also clustered in junctional domains (Figure 23). But in contrast to skeletal muscle, the large intramembrane particles that represent the DHPRs are not grouped into tetrads and bear no apparent relationship to the position of the feet [154].

The lack of tetrads indicates that a direct molecular interaction between DHPRs and RyRs may not be possible in cardiac muscle. However, the concentration of DHPRs in the junctional domain opposite to the RyRs is consistent with a Ca^{2+}-induced Ca^{2+} release mechanism for cardiac e-c coupling under the local control of the L-type Ca^{2+} current. A role for Ca^{2+}-induced Ca^{2+} release, supplementary to the direct molecular coupling, has also been proposed for skeletal muscle e-c coupling [155]. The alternating position of coupled DHPR/RyR pairs and uncoupled RyRs could provide the structural bases for such a combined Ca^{2+} release process. In this case, the initial Ca^{2+} release could be achieved by the Ca^{2+} release channels, linked to the voltage sensor DHPR, while the uncoupled release channels could provide an amplification mechanism, gated by the Ca^{2+} released by the coupled channels.

However, the steep dependence of SR Ca^{2+} release on surface membrane voltage in skeletal muscle does not necessarily need such an amplification [156]. In conclusion, the structural and functional properties of the Ca^{2+} release apparatus in skeletal and cardiac muscle fit nicely into the current hypotheses of cardiac and skeletal e-c coupling. Verification of these hypotheses, however, requires the identification of the molecular interactions of DHPR and RyR, that underlie their coordinated spatial organization in the junctions.

Molecular Bases of DHPR-RyR Interactions

The bond between surface membranes and SR is strong enough that the junctions survive the strains of muscle contraction as well as the stringent fractionation procedures used in the isolation of triads [157]. Even after forced dissociation, junctional T tubule and SR vesicles are capable of reassociating with one another. The cytoplasmic loop between repeats II and III of the a, subunit of the DHPR is a critical determinant of the specific mode of DHPR-RyR interactions in skeletal and cardiac e-c coupling. When chimeras of the skeletal and cardiac isoform of the a subunit were expressed in dysgenic myotubes, e-c coupling was restored, and the coupling mode was determined solely by the origin (skeletal or cardiac) of the II-III loop [159, 160]. The importance of the II-III loop of the DHPRs is emphasized by the recent observations that peptides from this portion of the molecule affect the properties of the reconstituted channels [161] and influence Ca^{2+} release from SR vesicles [162]. Interestingly, not only the peptide derived from the skeletal DHPR, but also that of cardiac origin interacted with the skeletal RyR, while neither had an effect on the cardiac isoform. This was unexpected, considering the results of the above mentioned expression study with skeletal-cardiac chimera. It suggests that the structural difference in the association of feet and DHPR particles [154] and the functional difference in the properties of e-c coupling in skeletal and cardiac muscle are at least partially determined by the RyR isoform. However, the interactions between DHPRs and RyRs in skeletal muscle need not be unidirectional. Recent evidence from a RyR deficient skeletal muscle cell line indicates that the lack of the RyR reduces the current density, but not the charge movement associated with the DHPR. Thus, the DHPR-RyR interactions may also be essential to DHPRs' function as Ca^{2+} channels [163]. Since the affinity of DHPRs for RyRs is low, the ability of the two proteins to remain associated with each other in skeletal muscle triads may reflect the concerted action of numerous DHPR/RyR pairs. Alternatively, interactions between the two channels may depend on the presence of an intermediate protein [164]. A 95-kDa protein (triadin), showing affinity for both DHPRs and RyRs, has been isolated from the junctional SR [165]. Triadin appears simultaneously with DHPRs and RyRs during development and is expressed in skeletal and cardiac muscle [166-169]. Thus, triadin is a possible candidate for linking the two major junctional proteins. However, structural predictions from the amino acid sequence suggest that triadin may be mostly positioned on the luminal and not the junctional side of the SR membrane [168]. In light of the close proximity of DHPRs and RyRs and of their functional interaction, it has been assumed that these two components also play a major role in the formation and maintenance of the junctions. Junction formation is much more complex and that although DHPRs and RyRs are clearly the functionally significant components of the junction.

Clusters of DHPRs and RyRs Form Concomitantly

Several lines of evidence indicate that during normal development the molecular differentiation of the apposed junctional domains of surface membranes and of the SR proceed concomitantly. In cultured skeletal myotubes, the first ultrastructurally identifiable junctions appear at the same time as the first DHPR/RyR clusters, and clusters of either channel are not seen independently of each other [171].

The observations rather support a mechanism of a concurrent accrual of the two channels in the newly formed and expanding junctions. However, based on the above information alone, it cannot be decided whether the formation of a junction and the accrual of the two channels are two aspects of the same molecular mechanism or represent two distinct processes.

Targeting of DHPRs and RyRs

In this section are considered two questions: how do RyRs and DHPRs reach the junction, and how are the two proteins targeted to peripheral versus internal junctions? Figure 24 illustrates two possible pathways by which RyRs and DHPRs could be transported from their site of synthesis in the rough endoplasmic reticulum to the junctions, either independently or jointly.

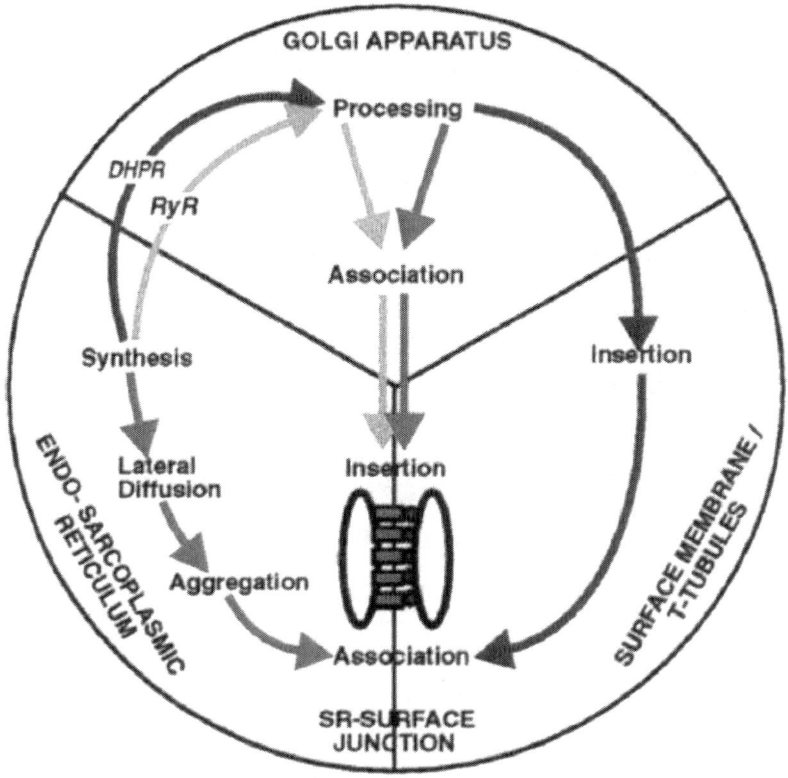

Figure 24. Alternative transport pathways of the DHPR and RyR [484]. After synthesis in the rough endoplasmic reticulum, DHPRs and RyRs could be shuttled through the Golgi apparatus for further processing and then transported to the surface membrane and sarcoplasmic reticulum, respectively (light gray). Their association with one another could occur before or after insertion of the transport vesicles into the target organelles. Alternatively, RyRs could be targeted to the junctional domain by lateral diffusion in the continuous endo-sarcoplasmic reticulum and subsequently associate with DHPRs, which reach the plasma membrane or T tubules by the constitutive exocytic pathway (dark gray).

According to the first hypothesis, RyRs could diffuse from the rough endoplasmic reticulum directly into the SR; in developing muscle, the two membranes are indeed continuous with each other. DHPRs could follow the exocytic route through the Golgi *apparatus* and be inserted into the surface membrane by constitutive fusion of exocytic vesicles. In this scheme, DHPRs and RyRs would reach the junctions independently of each other. An alternate hypothesis of junction formation has been proposed based on the distribution of DHPRs, RyRs, and a T tubule antigen (TS28) in developing rabbit muscle [170]. Colocalized clusters of RyRs and DHPRs were observed either at the fiber periphery or in proximity of TS28-labeled membranes. These represent peripheral couplings and triads that are known to coexist in developing myotubes and muscle fibers. However, DHPR/RyR clusters were also noted at some distance from the plasma membrane and from TS28-positive membranes. This has been interpreted to indicate the formation of junctions between Golgi-derived DHPR- and RyR-bearing vesicles before their fusion with the SR and either plasma membrane or T tubules. Consequently, fusion of the preformed junctions with the plasma membrane or T tubules, and with the SR had to occur in a highly coordinated manner. Several observations are inconsistent with this hypothesis.

First, using similar antibody combinations in myotubes developing in culture, internal DHPR/RyR clusters were never found independent of a general marker for the T-system, but they were expressed simultaneously with or after the formation of T tubules [174, 175].

Second, a model in which preformed junctions are inserted into the two membrane systems is inconsistent with observations of junctions without feet in normal and mutated muscle (see below), unless two independent mechanisms of membrane attachment exist-one for the DHPR and RyR vesicles and another for the undifferentiated membrane systems.

Third, a quantitative study of junction formation in cardiac muscle [172] supports a mechanism of continuous growth of junctional area occupied by DHPRs and RyRs rather than a mechanism whereby quantal growth of junctions would be expected.

An interesting observation related to the targeting of DHPRs is the successive location of this protein first in junctional domains of the plasma membrane and later during development in junctional T tubules [170, 176,177]. Few skeletal muscles make do without T tubules and simply use peripheral junctions between SR and plasma membrane for e-c coupling. Those muscles that do have T tubules temporarily establish a peripheral Ca^{2+} release system during early development, which is morphologically, physiologically, and molecularly equivalent to the internal triads [170, 171, 177, 178]. However, as soon as a T-system develops, internal junctions between it and the SR are formed and peripheral junctions are eliminated [177]. Thus most muscles must have a secondary targeting mechanism, which is activated in parallel to T tubule development and is responsible for the location of DHPR clusters in the T tubules and not in the sarcolemma. Regardless of whether RyRs and DHPRs find their way to the junctions independently of each other or in associated clusters, the basic questions of how DHPRs and RyRs associate with one another, how the initial SR-surface/T tubule junctions are formed during development, and how junctions are targeted to T tubules are still largely unanswered. Some information derived from mutant model systems is dicussed below.

Triad Formation without DHPRs

In the absence of compelling evidence for a direct binding of isolated DHPRs and RyRs, investigations have concentrated on the potential contributions of the individual channels to triad formation. Valuable information was obtained from the study of muscular dysgenesis, which is a recessive lethal mutation with a deficiency in skeletal muscle e-c coupling [135, 179].

Myotubes from the homozygous embryos lack the DHP-sensitive, L-type Ca^{2+} currents [135, 180] and the charge movement associated with e-c coupling [135]. These properties are restored by transfection with a plasmid carrying the cDNA for the a, subunit of skeletal DHPR [181, 182]. The primary defect of muscular dysgenesis has been traced to a single point mutation, resulting in a shift in the reading frame and an unstable message for the a, subunit [183]. The other subunits of DHPRs (a2, j, and y) are expressed [184], but a2 is not targeted correctly to the T tubules (47), suggesting that the a, subunit is required for the normal incorporation of a2 into junctional membrane domains.

But how does the lack of the DHPR a subunits affect the formation of junctions with the SR and the normal organization of RyRs within these junctions? T tubules and SR develop normally in dysgenic myotubes in culture [185], but triads with visible feet seemed to be lacking [186, 187]. Immunocytochemistry showed failure of RyRs and triadin to cluster [167].

The normal coclustered organization of DHPRs, RyRs, and triadin could be restored by *de novo* expression of the a, subunit in segments of dysgenic myotubes rescued by fusion with normal rat fibroblasts. These findings could have been interpreted in support of a requirement of DHPRs in the clustering of RyRs at junctions.

Furthermore, they showed that a small number of dysgenic myotubes have been observed in the absence of skeletal a1 subunits, that formed in normally distributed RyR clusters. The ability of RyRs to cluster and form junctions in the absence of DHPRs was further confirmed *in vivo*, by the observation that triads with well-ordered arrays of feet are present in the diaphragm of dysgenic mouse embryos, a muscle that develops fairly well despite the lack of e-c coupling [152].

Interestingly, colocalization of triadin and RyR *in vitro* was restricted to those few myotubes, where RyRs were clustered. The distribution pattern of triadin was different from that of the RyRs, wherever the latter were diffusely distributed.

These findings suggest an interdependence of RyRs and triadin in the junction. On the other hand, clustering of RyRs in ordered arrays, association of triadin with them, and the formation of junctions between SR and T tubules seem to be independent of the presence of skeletal type DHPRs. This, however, did not entirely rule out the possibility that a different a, DHPR isoform, which is not recognized by the skeletal isoform-specific antibody, is responsible for junction formation in dysgenic myotubes. Studies demonstrating transient expression of the cardiac isoform during myogenesis [188, 189] and recordings of DHP-sensitive Ca^{2+} currents with cardiac characteristics from dysgenic myotubes substantiated these concerns [190]. The question could be addressed in a new cell line of dysgenic muscle origin, that formed RyR clusters at high frequency, but still did not contract [191]. These RyR clusters corresponded to T tubule-SR junctions with ordered arrays of feet, as seen in electron microscopy, but did not contain al and a2 subunits of the DHPR. The absence of the a2 subunit, which associates with all known a isoforms, makes it rather unlikely that another a, subunit substituted for the missing skeletal isoform's role in junction formation, but not in e-c coupling. Thus, accumulating evidence indicates that despite the close structural association of DHPRs and RyR in skeletal and cardiac junctions, the DHPR is not needed for clustering of RyRs or for junction formation.

Formation of Ordered Arrays Is an Inherent Property of RyRs

The tendency of RyRs to form ordered arrays in junctions and the apparent absence of RyR aggregates outside of junctions in vertebrate skeletal muscle, raise the question whether the characteristic disposition of RyRs is dependent on other junctional components. In cardiac muscle and in some skeletal muscles of invertebrates, ordered arrays of feet are seen on SR surfaces that are not associated with T tubules or the plasma membrane [192, 193]. Immunocytochemistry confirmed that feet in these regions called extended or corbular SR, actually represent RyRs [169, 194, 195]. Thus, the formation of a junction may facilitate the organization of RyRs at these sites, but is not needed for the formation of feet arrays. Even more so, it appears that muscle-specific components of the SR are not required at all for the regular organization of RyR. RyRs expressed in Chinese hamster ovary cells become correctly inserted into the endoplasmic reticulum. Their feet face the cytoplasm and form extensive arrays with the same periodicity as in skeletal muscle [158].

Finally, even purified RyRs possess the potential for self-assembly into arrays, very much like those in muscle, despite the absence of other proteins.Thus the organization of RyRs into ordered arrays can occur independently from interactions with other components of either the SR or the surface membrane and probably represents an intrinsic property of the protein. The array is most likely formed by interactions of the cytoplasmic domains of RyRs (the feet) with each other [149]. Some external factors, however, may be needed to specify the location of these arrays in the skeletal muscle junctions and at specific sites in the sarcomeres of cardiac and invertebrate muscles or else to prohibit arrays from forming in large portions of non-junctional SR.

Triad Formation without RyRs

While interactions between RyRs and DHPRs are apparently not required for junction formation between SR and surface membrane, binding of RyRs with an unknown component of the surface membrane could still be involved in the docking of the two membrane compartments. This question was explored in developing muscle fibers from a transgenic mouse with a targeted disruption of the skeletal muscle RyR gene [196]. The phenotype of this knock-out mouse is very much like that of the dysgenic mutant. Skeletal muscle fibers initially develop normally, but remain paralyzed due to a defect of e-c coupling. During later embryonic development, this caused retardation of muscle development and the mice died at birth due to respiratory failure. Northern blot analysis verified that the message of the skelctal muscle RyR was specifically missing and physiological analysis showed that depolarization failed to induce Ca^{2+} release. However, myotubes from the RyR knock-out mouse still responded weakly to caffeine with a release of Ca^{2+} from the intracellular stores, indicating the presence of small numbers of RyRs, later shown to be of the neuronal type [197].

How does the lack of skeletal RyR affect the ability of the muscle cell to form junctions between SR and plasma membrane or T tubules, and how does it affect the organization of the DHPRs within the junctional domains? Surprisingly, triads and peripheral couplings formed in the RyR knockout mouse, even though the junctions did not contain feet [198]. For this phenotype, the RyR mutant was named "dyspedic" mouse. The failure of e-c coupling in dyspedic mice and the lack of feet in their SR/surface junctions confirms that the disrupted skeletal muscle RyR gene is responsible for depolarizationinduced Ca^{2+} release and that it actually gives rise to the feet.

However, feet are apparently not required for the docking of the SR to the surface membrane. The width of the junctional gap in dyspedic muscle is irregular and narrower than in normal controls, indicating that in normal junctions, the feet determine the distance between SR and surface membrane.Therefore, it confirmes that the feet span the whole distance of the junctional gap.

Whereas RyRs are not required for the formation of the junction, they may be involved in the specific organization of the DHPRs in the junctions. Dyspedic muscle *in vivo* expresses DHPRs, although at a level much lower than that of normal fibers [197]. In these muscles, the large intramembrane particles representative of DHPRs were neither found in tetrads, as in normal skeletal muscle, nor randomly clustered in the junctional domain, as in cardiac muscle [198]. In contrast, primary myotubes derived from a second dyspedic knockout revealed L-type current densities, similar to those of normal myotubes [163] and in a cell line derived from the same mice the DHPR particles, were clustered in the junctional domains of the plasma membrane.

Thus, DHPRs apparently require the presence of RyR arrays for their own arrangement into tetrads. On the other hand, clustering of DHPRs at the junctions occurs in the absence of RyRs, but may be dependent on expression levels. The random disposition of DHPRs in the junctional domains of cardiac muscle confirms that the DHPRs are capable of aggregating in the junctions, despite the lack of a close spatial relationship with the RyRs.

It is worth mentioning that during development, the feet appear coordinated with other ultrastructural characteristics of the juncton. The electron-dense content of the terminal SR cisternae, presumably calsequestrin, follows closely the differentiation of the feet in the junction. Also, the T tubules frequently show some electron-dense content, most likely components of the extracellular matrix, specifically in the junctional area.

Thus, the mechanism of junction formation may involve transmembrane complexes in the SR and in the surface membrane. Together they make up an extensive junctional complex including extracellular matrix, integral membrane proteins of the surface membrane and the SR, cytoplasmic components between the membrane compartments, and luminal SR proteins. Each of these may contribute in some way to the formation of the functional junction.

Docking: An Initial Step in Junction Formation

The capability of junctions to form in the absence of either DHPRs or RyRs, and perhaps also in the absence of both, suggests that during normal development, docking of SR to the surface membrane precedes the clustering of feet and their association with the DHPR particles. This is supported by reports of peripheral couplings entirely or partially without feet in normal developing muscle [173, 199] and of similar SR-surface junctions in a myogenic nonfusing cell line (BC3H1), before expression of DHPRs and RyRs [200]. Closer analysis of the BC3H cells revealed that cultures with clustered and colocalized DHPRs and RyRs contained junctions, that were only partially occupied by feet as well as plasma membrane patches, that were only partially occupied by tetrads [172]. Further evidence for a gradual accumulation of RyR in newly formed junctions comes from the ultrastructural analysis of developing avian cardiac muscle [154]. In these muscle fibers, junctions without feet appeared first, and their frequency decreased in parallel, with the gradual increase in junctions that were either partially or totally occupied by feet. Furthermore, the surface areas of junctional domains occupied by feet and by DHPRs increased precisely in parallel to each other, indicative of a coordinated accrual of the two proteins in the developing junctions. Taken together, the above information allows the proposal of a new model for the formation and differentiation of triads and related junctions, with three discrete steps (Figure 25): (1) the docking of a patch of SR to a corresponding patch of surface membrane; (2) the accumulation of RyRs in the junction and their organization into arrays; and (3) the accumulation of DHPRs in the junctional domain of the surface membrane.

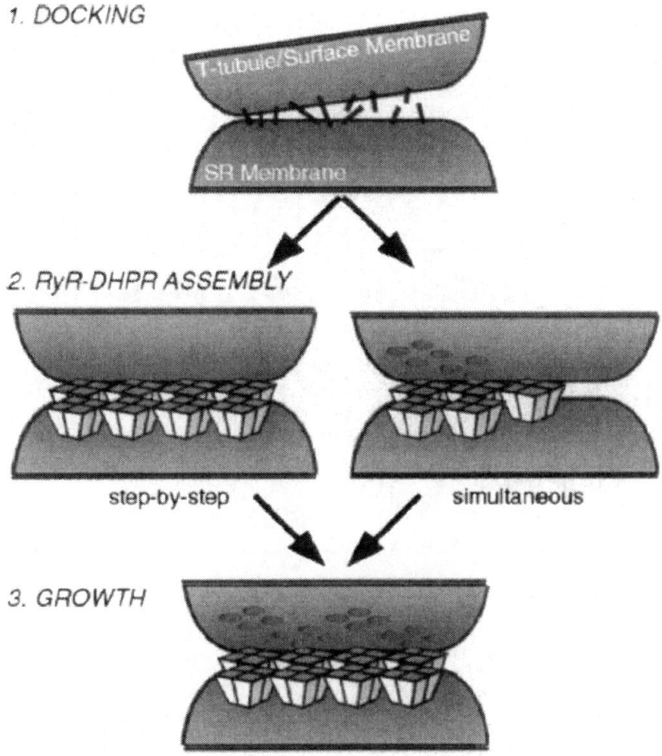

Figure 25. Three stages of junction formation of the skeletal muscle e-c coupling apparatus [484]. A schematic view into the junctional gap shows the hierarchy of events leading to the differentiation of the SR-surface membrane junction. As of now unidentified components induce the docking of undifferentiated patches of SR with the surface membrane. Subsequently arrays of feet (RyR, squares) self-assemble within the junctional domain. This determines the width of the junctional gap and provides a scaffold for the regular assembly of tetrads (DHPRs, circles) opposite every other foot. The assembly of tetrads may occur as an individual step or it may sequentially follow the assembly of the feet arrays. After the first regions of the junction have been formed and have become functional, the junction continues to grow by the accrual of additional RyRs and DHPRs.

Skeletal muscle DHPRs associate with RyRs forming tetrads, while cardiac muscle DHPRs become aggregated opposite the RyR arrays, but form only loose and irregular associations with the RyRs. The three steps are to be understood as a hierarchy of events, where docking determines the site of RyR aggregation, and the RyR arrays cause the accumulation and tissue-specific organization of the DHPRs. Whereas docking and the incorporation of the channels can be temporarily dissociated, the initial aggregation of RyRs and DHPRs occurs virtually simultaneously. Further gradual accretion of RyRs, closely followed by the DHPRs, occurs as the junction matures.

Developmental and Tissue-Specific Regulation of Skeletal and Cardiac Muscle Calcium Channels Involved in Excitation-Contraction Coupling

It has been shown that fetal development in utero and myogenic differentiation are abnormal in the dysgenic (mdg) [201] mouse, which does not express the α_1 subunit of the DHPR. Moreover, this mutation is lethal at birth [202, 203]. Thus, regulation of calcium channel gene expression in early stages of development is of interest in terms of the effects of altering calcium homeostasis on gene expression, whereas in the later developmental stages, the quality and quantity of muscle contraction are directly influenced by channel expression. The five subunits of the DHPR have been termed α_1, α_2, β, δ, and γ (the α_2 and δ subunits are encoded by a single gene and are linked by disulfide bonds) [204, 205]. The DHPR subunits have been characterized structurally, [206-209] and recent studies have begun to define their functional roles. The α_2 subunit modulates activation of the α_1 subunit, [205] and the β subunit substantially accelerates the activation and inactivation kinetics of the calcium channel current [210, 211]. Numerous studies have demonstrated functional differences between fetal and mature hearts in terms of calcium handling [212-217]. Moreover, it has been shown that newborn hearts require higher extracellular calcium to optimize contractility, and a "deficiency" in sarcoplasmic reticulum development has been proposed [218-221]. Furthermore, it has been shown that the neonatal heart is highly sensitive to calcium channel blockers [222].

Both, skeletal and cardiac muscle expression of the calcium channels, required for E-C coupling is developmentally upregulated in the rabbit [485]. In all cases, the day of birth is marked by a consistent 2-fold increase in calcium channel mRNAs (Figure 26 and 27).

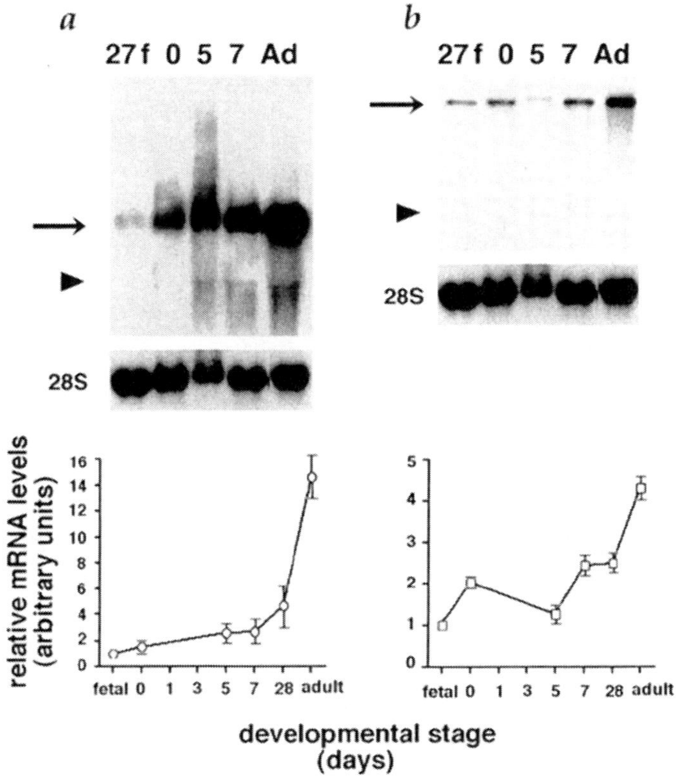

Figure 26. Developmental expression of dihydropyridine receptor (DHPR) and ryanodine receptor (RyR) mRNAs in skeletal muscle [485]. A representative Northern blot hybridized separately to DHPSKEL1 (a) and CRC105 (b). Both the 6.5 kb DHPR a1 subunit mRNA (a) and the 16 kb RyR mRNA (b) increased in relative abundance throughout muscle development. Time points are 27 days before birth (27f), day of birth (0), 5 days after birth (5), 7 days after birth (7), and adult (Ad). Each lane was loaded with 20 gg of total RNA. Hybridization to the 28S ribosomal RNA is shown below each Northern blot. Quantification of normalized DHPR a, subunit and RyR mRNA expression levels during skeletal muscle development demonstrates calcium channel mRNA levels expressed relative to fetal values in arbitrary units (shown in the graph below each representative Northern blot). Each time point represents values determined from RNA samples from three or more animals. Error bars represent SEM.

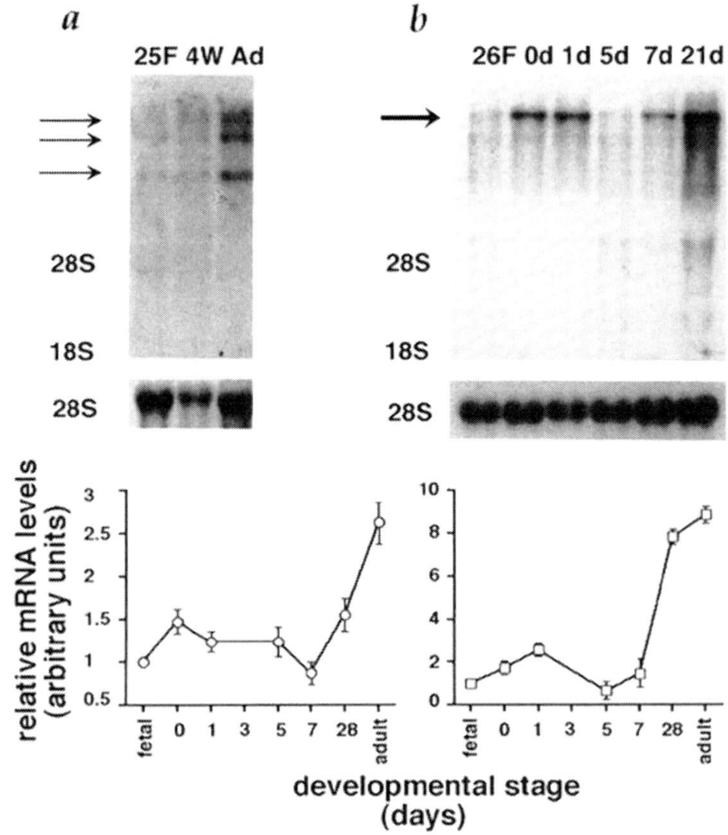

Figure 27. Developmental expression of dihydropyridine receptor (DHPR) a1 subunit and ryanodine receptor (RyR) mRNAs in cardiac muscle [485]. a, A representative Northern blot analysis of DHPR a1 subunit mRNA during cardiac development is shown. Three DHPR mRNA species (arrows) are shown at 25-day fetal (25F), 4-week postnatal (4W), and adult (Ad) time points. b, A representative Northern blot analysis of RyR mRNA during cardiac development is shown at 26-day fetal (26F) and 0-, 1-, 5-, 7-, and 21-day postnatal (Od, 1d, 5d, 7d, and 21d, respectively) time points. Each lane was loaded with 20 gg of total RNA. Hybridization to the 28S ribosomal RNA is shown below each Northern blot. Quantification of normalized mRNA levels is shown in the graph below each representative Northern blot. Each time point represents values determined from RNA samples from three or more animals. Error bars represent SEM.

The signal for this increased expression has not been identified, but could be related to increased physiological stresses on both the skeletal and cardiac muscle systems experienced during birth, including the conversion from fetal to neonatal circulation in the cardiovascular system. All four calcium channels in both forms of striated muscle exhibited a steady upregulation of expression (measured in terms of mRNA levels) during development. There were corresponding increases in specific [5H] ryanodine binding in both cardiac and skeletal muscles.

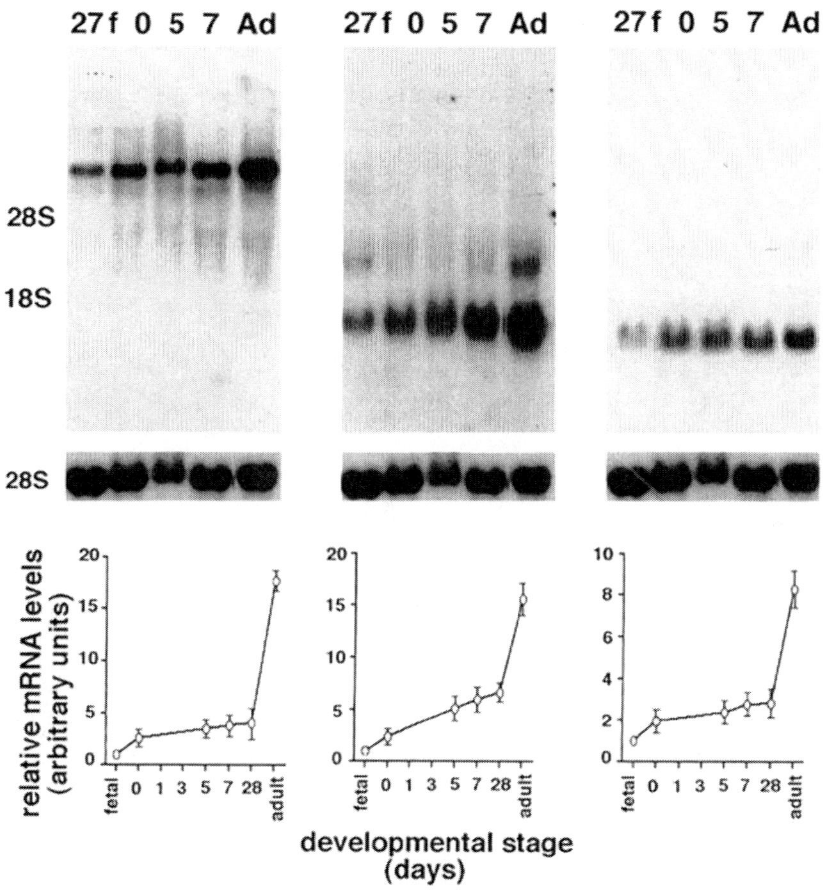

Figure 28. Developmental expression of skeletal muscle dihydropyridine receptor (DHPR) α_2, β, and γ subunits [485]. Northern blot analyses of α_2 subunit (a), β subunit (b), and γ subunit (c) expression during fetal, neonatal, and adult developmental stages. Time points are 27 days before birth (27f), day of birth (0), 5 days after birth (5), 7 days after birth (7), and adult (Ad). Each lane was loaded with 20 gg of total RNA. Hybridization to the 28S ribosomal RNA is shown below each Northern blot. Quantification of normalized mRNA levels is shown in the graph below each representative Northern blot. Each time point represents values determined from RNA samples from three or more animals. Error bars represent SEM.

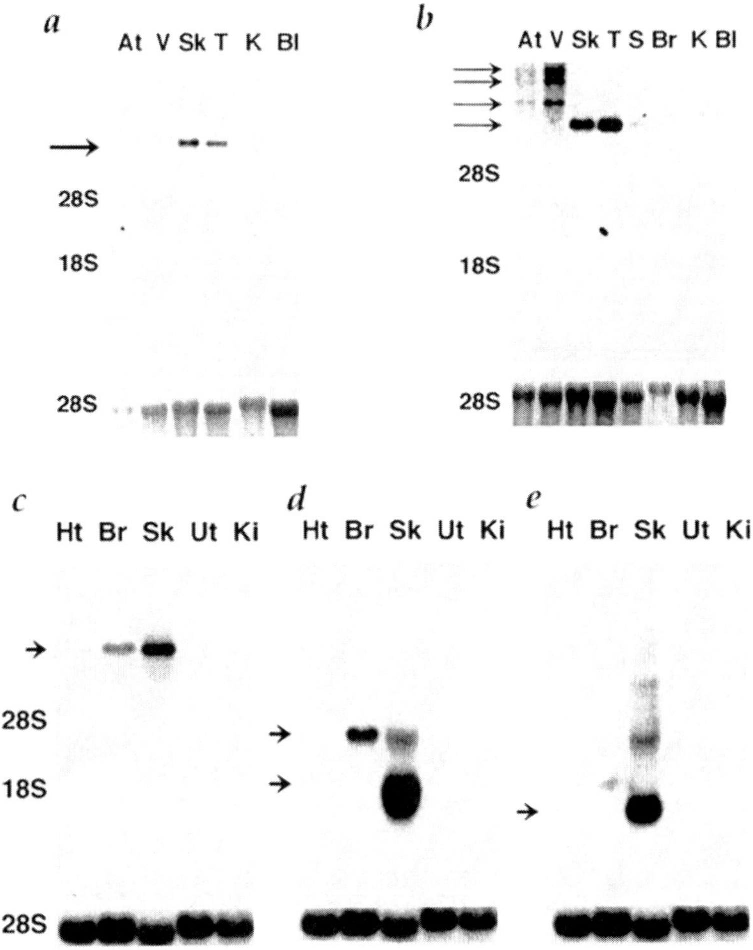

Figure 29. Northern blot analyses demonstrating specificity and tissue distribution of skeletal and cardiac muscle dihydropyridine receptor (DHPR) cDNA probes [485]. a, Tissue distribution of skeletal muscle DHPR a, subunit mRNA is shown. DHPSKEL1 hybridized to a single 6.5 kb mRNA species (arrow) in total skeletal muscle (Sk) and fast-twitch tibialis anterior muscle (T). No hybridization signals were detected in atrium (At), ventricle (V), kidney (K), or bladder (Bl). b, Tissue distribution of cardiac DHPR a1 subunit mRNA is shown. DHPCARD1 recognized both the cardiac and skeletal isoforms of the a1 subunit of the slow calcium channel. In both atrium (At) and ventricle (V), three mRNA species were detected (top three arrows). In addition, the cDNA probe hybridized to the 6.5 kb skeletal muscle DHPR isoform (bottom arrow) in mixed skeletal muscle (Sk), fast-twitch tibialis anterior muscle (T), and slow-twitch soleus muscle (S). No signals were detected in brain (Br), kidney (K), or bladder (BJ). c, Tissue distribution of skeletal muscle DHPR a2 subunit is shown. (Continued on next page).

The a2 subunit cDNA detected a single 8 kb mRNA species (arrow) in heart (Ht), brain (Br), and skeletal muscle (Sk), in increasing relative amounts. Longer exposure also detected the -8 kb mRNA in uterus (Ut) and kidney (Ki) (not shown). d, Tissue distribution of skeletal muscle DHPR, subunit is shown. The p subunit cDNA hybridized to a single 3.4 kb brain mRNA species (Br, top arrow) and to both the 3.4 kb brain isoform and a 1.8 kb skeletal isoform in mixed skeletal muscle (Sk, bottom arrow). No signals were detected in heart (Ht), uterus (Ut), or kidney (Ki). e, Tissue distribution of skeletal muscle DHPR y subunit is shown. The y subunit cDNA probe recognized a major 1.2 kb mRNA in skeletal muscle (Sk); no signals were detected in heart (Ht), brain (Br), uterus (Ut), or kidney (Ki). In all lanes, 20 gg of rabbit total RNA was loaded. The positions of the 28S and 18S ribosomal RNAs are denoted as size markers. Hybridization to the 28S ribosomal RNA is shown below each Northern blot.

These increases in cardiac and skeletal muscle calcium channel expression support the findings of earlier functional studies that reported a "maturation" of sarcoplasmic reticulum during development. However, previous functional studies did not identify specific molecular events associated with developmental changes in sarcoplasmic reticulum function.

The present study establishes, that the calcium channels required for E-C coupling are in fact upregulated at important developmental stages and provides a basis for understanding alterations in calcium fluxes observed, particularly when comparing fetal with adult muscles. The present study is in agreement with a previous study that examined the expression of sarcoplasmic reticulum proteins during development [223] and extends that study to now include the other calcium channel responsible for E-C coupling, the DHPR and its subunits. The expression of each of the skeletal muscle DHPR subunit mRNAs were coordinately regulated during development (Figure 28a through 28c).

This finding suggests that the relative levels of each of the DHPR subunits remain comparable throughout development, although the increases in γ subunit expression are less pronounced than those observed for both the α_2, and β subunits. It has been speculated that alterations in calcium channel function observed during development could be due to changes in subunit expression or in the expression of different isoforms of the DHPR subunits. In this regard, it is interesting to note that the predominant form of the DHPR β subunit identified by Northern analysis in the brain is a 3.4 kb species (Figure 29d).

This larger mRNA is detected in skeletal muscle, but the major form is the 1.3 kb mRNA (Figure 29d). During development, the brain form is expressed in fetal tissue, but its levels decrease during the early postnatal period and subsequently rise again in the adult (Figure 28b). It has not been established whether these two mRNAs encode functionally distinct β subunits. If so, they could interact with the same DHPR α_1 subunit(s), but invoke different effects, which might be developmentally regulated.

On the day of birth, when the heart is first exposed to increased pressures associated with switching from fetal to neonatal circulation, calcium channel mRNA levels increased by 2-fold. Growth factors could be involved in regulating these alterations in ion channel expression. For example, transforming growth factor- β has been shown to upregulate RyR expression [225], whereas other growth factors, most notably fibroblast growth factor, downregulate calcium channel gene expression [226].

The pattern of cardiac calcium channel gene regulation during development is reminiscent of that observed for cardiac-specific genes during heart failure. It has been reported that during heart failure fetal isoforms of cardiac-specific genes are expressed [227].

This reversion to the fetal phenotype has been shown to be mediated by growth factors [228]. Previously it was reported that DHPR57 and RyR4' expression are decreased in end stage human heart failure. The decreased expression of calcium channels in human heart failure simulates the fetal pattern of expression of calcium channel mRNA, observed in the rabbit. The present study provides a molecular basis for understanding the reported sensitivity of neonates to calcium channel blockers [229]. Our data would suggest that at fetal and neonatal stages, when expression of calcium channels involved in E-C coupling is low, small doses of channel blockers could have adverse effects on the myocardium and on pulmonary performance. Treatment with calcium channel blockers could result in diminished cardiac contractility, vascular tone, and diaphragmatic activation. This heightened sensitivity to calcium channel blockers in fetal and neonatal tissues is offset by the protective effect of reduced cellular calcium influx during hypoxia and acidosis.

Chapter 3

Roles of the Sarcoplasmic Reticulum

Abstract

The governor of a gasoline or diesel engine provides a feedback mechanism to change speed as needed and to maintain a specified speed once it is reached. In an analogous manner, the sarcoplasmic reticulum (SR) functions as a dynamic calcium governor in muscle, that provides automatic feedback control for altering and maintaining myoplasmic and SR calcium levels. Interactions between terminal regions of the SR and the transverse tubular membrane, as well as the envelopment of the myofibrils by the longitudinal SR, uniquely position this organelle for controlling calcium release and reuptake during muscle contraction. Given its central role in controlling calcium cycling, the SR has developed an elaborate set of calcium-regulatory proteins that provide for a near-instantaneous release of calcium upon excitation and a slower means of calcium reuptake and maintained calcium storage during muscle relaxation and inactivity. Specifically, the processes of calcium storage, release, and reuptake are balanced by the concerted action of three major classes of SR calcium-regulatory proteins: (1) luminal calcium-binding proteins (calsequestrin, histidine-rich calcium-binding protein, junctate, and sarcalumenin) for calcium storage; (2) SR calcium release channels (RyR and IP_3 receptors) for calcium release; and (3) sarco(endo)plasmic reticulum Ca^{2+}-ATPase (SERCA) pumps for calcium reuptake.

The Actors on the "Sarcoplasmic Reticulum" Scene

The SR is the primary calcium-storage organelle in striated muscle. The efficacy of calcium release during EC coupling depends on the level and calcium-binding capacity of proteins within the SR. A number of different acidic calciumbinding proteins participate in calcium buffering and help to maintain SR free calcium at levels around 1 mM [230] The most important protein in buffering and storing calcium in the terminal SR is calsequestrin (Figure 30A), which comprises 27% of all junctional SR protein [231].

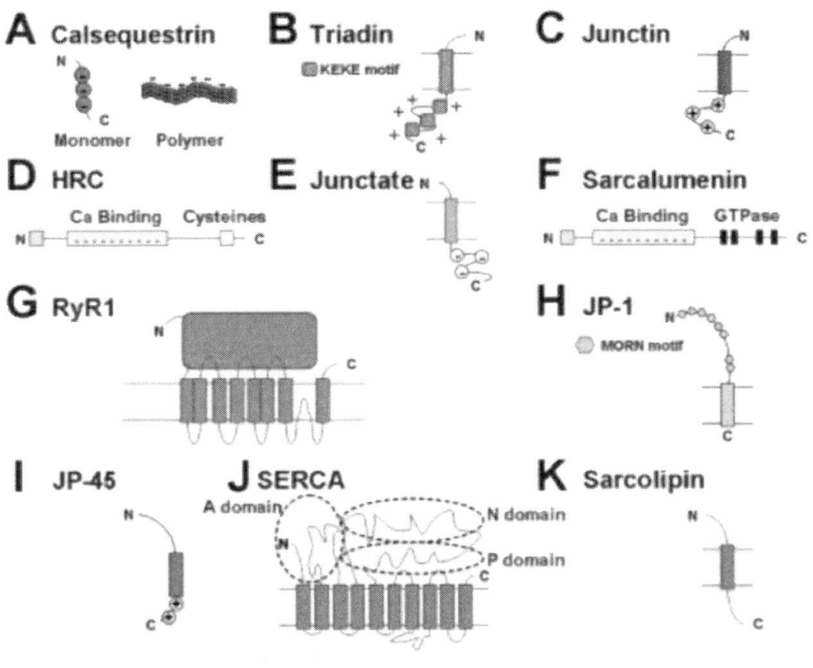

Figure 30. Structural features of key calcium-regulatory proteins of the skeletal muscle sarcoplasmic reticulum [486]. (A) Calsequestrin. (B) Triadin. (C) Junctin. (D) Histidine-rich calcium binding protein (HRC). (E) Junctate. (F) Sarcalumenin. (G) Ryanodine receptor (RyR1). (H) Junctophilin type 1 (JP-1). (I) Junctophilin 45 (JP-45). (J) SR/ER Ca^{2+}-ATPase (SERCA). (K) Sarcolipin.

Although originally regarded as only a luminal calcium buffer, calsequestrin is now recognized to play a more direct role in regulating SR calcium release during skeletal muscle e-c coupling [232] Calsequestrin was first extracted from SR vesicles in the early 1970s by MacLennan and colleagues [233]. Initial studies found that the protein is highly acidic, hydrophobically bonded to the inside of SR vesicles, and capable of low-affinity (kD 40 M) and high capacity calcium binding (40 mol Ca^{2+}/mol protein at pH 7.5). Thus, the protein was given the name "calsequestrin" for its ability to sequester calcium within the SR [234]. Since its original isolation, two isoforms of calsequestrin (skeletal and cardiac) have been identified, which differ in the number of acidic residues in their C-termini. Although, it bears no effect on ability to bind calcium, skeletal calsequestrin has fewer acidic residues in its C-terminal tail than the cardiac isoform [232]. Fast- and slow-twitch skeletal muscle express the skeletal isoform [235], whereas the cardiac isoform is expressed in both cardiac muscle and slow twitch skeletal muscle. Interestingly, both isoforms of calsequestrin are present early during fast-twitch skeletal muscle development. As development progresses, expression of the cardiac isoform gradually decreases until it is essentially absent in adult muscle [236]. At the protein level, the switch from the cardiac to skeletal muscle isoform occurs during postnatal weeks 2– 4 [237]. Synthesis of calsequestrin is controlled by both neuronal input [238] and myogenin [239], a member of the muscle regulatory factor (MRF) family of muscle specific transcription factors. Once synthesized, skeletal muscle calsequestrin becomes localized in the lumen of the terminal SR by an as yet unknown mechanism, given that it does not possess a C-terminal Lys-Asp-Glu-Leu (KDEL) ER retention signal characteristic of other soluble SR proteins [232]. Calsequestrin has three similar domains linked by short unstructured sequences of amino acids to form a monomer with a hydrophilic core. Each domain has a disk-like shape with a four-stranded-sheet at its hydrophobic core and two hydrophilic-helices on each side of the sheet. The structure of each domain is reminiscent of *Esherichia coli* thioredoxin. At turns between the -strands, hydrophilic side chains form crevices that are similar to the redox centers of thioredoxin [240]. This similarity to thioredoxin structure suggests that calsequestrin may participate in redox reactions by acting as an SR chaperone protein, that mediates disulfide bond formation in oxidative protein folding [232]. Approximately one-third of calsequestrin is composed of acidic residues, most of which are in the C-terminus of the monomer. These acidic residues are important for calcium binding. Rather than the presence of discrete calcium-binding sites, pairs of acidic residues create a net surface charge to which calcium binds [240]. As calcium binds, the hydrating water is freed in an energetically favorable reaction [241]. Calcium concentrations in the low micromolar range (10 M) cause collapse of the monomer's structure.

Further increases in calcium (10 –100 M) lead to dimerization, which results in the formation of an electronegative pocket for enhanced calcium binding. Polymerization follows from dimerization and additional calcium binding at higher levels of calcium. At typical SR free calcium concentrations (1 mM), calsequestrin exists as a stable polymer [240]. Polymers of calsequestrin are visible as electrondense material in the terminal SR in electron micrographs of the skeletal muscle triad [242]. However, calsequestrin is not membrane bound [243], it resides in the lumen of the SR. Calsequestrin is anchored to the junctional face of the SR membrane through interactions with junctin, triadin, and RyR1 (Figure 31) [244].

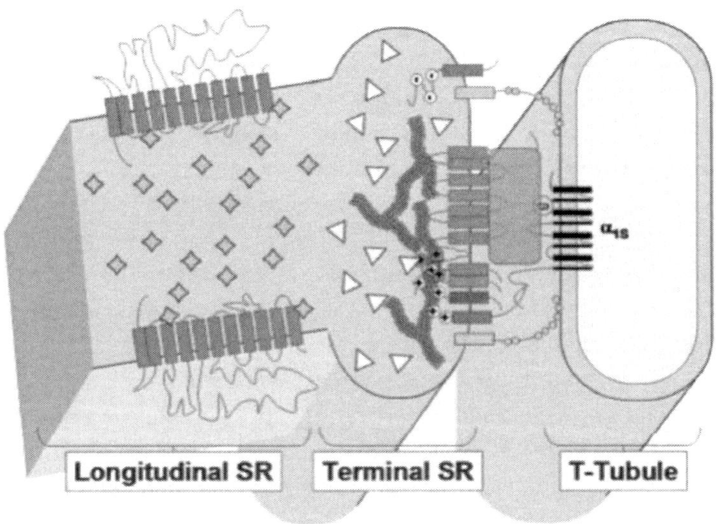

Figure 31. Location and interactions of the different SR calcium-regulatory proteins within one side of the triad [486]. Calsequestrin, triadin, junctin, junctate, RyR1, JP-1, JP-45, SERCA, and sarcolipin are depicted using the schematics shown in Figure 31. For simplicity, sarcalumenin and HRC proteins are shown as gray four-point stars and white triangles, respectively, and SERCA is shown only in the longitudinal SR. Potential interactions between RyR1 and the α_{1S}- and β_{1a}-subunits of the skeletal muscle DHPR, and the C-terminus of the α_{1S}-subunit and JP-45 are also shown.

Junctin and triadin binding involves an aspartate rich region (amino acids 354 –367) at the C-terminus of calsequestrin [245]. Potential direct interactions between calsequestrin and RyR1 have also been observed [246, 247]. Triadin was first identified in skeletal muscle (Figure 30B) [248]. Although triadin in skeletal muscle is primarily found as a 95-kDa glycoprotein embedded in the SR membrane, a number of different triadin splice variants have also been identified [249-251]. The bulk of the full-length triadin protein is located in the lumen of the SR [252], where it interacts with calsequestrin, junctin, and RyR1. Multiple Lys-GluLys-Glu (KEKE) motifs, located in the luminal portion of triadin are required for binding to calsequestrin95 and RyR1 [253]

Similarly, junctin is a 20-kDa protein with a short cytosolic N-terminus, followed by a single membrane-spanning -helix and a long, positively charged luminal C-terminus (Figure 30C). Like triadin, the luminal structure of junctin binds both calsequestrin and RyR1 [254] By binding to both RyR1 and calsequestrin, triadin and junctin anchor calsequestrin, and thereby increase calcium buffering close to sites of SR calcium release (Figure 31). In addition, the quaternary complex of triadin, junctin, RyR1, and calsequestrin is thought to relay information regarding SR calcium store content to RyR1 [255], ultimately affecting calcium release and, thus, skeletal muscle contractility. The means by which calsequestrin regulates the calcium-release mechanism is discussed in greater detail in the next section.

The role of histidine-rich Ca^{2+}-binding protein (HRC), as a second luminal calcium-storage protein, has recently been appreciated. In native SR, HRC (699 – 862 amino acids) exists as a multimer of five or more subunits [256]. The primary amino acid sequence consists of three regions: (1) a conserved N-terminal signal sequence; (2) a central acidic histidine-rich repeat region; and (3) a conserved C-terminal cysteine-rich region (Figure 30D) [257]. The central region is responsible for both calcium binding and interactions with other junctional proteins. Specifically, the concentration of acidic residues in the central region may provide charge on the protein necessary for calcium binding [258]. Nonetheless, HRC is less abundant in skeletal muscle than calsequestrin, indicating that it functions only as a secondary calcium-binding protein [259]. HRC also binds to triadin, perhaps signifying a greater role for this molecule than simply increasing the number of calcium binding sites in the SR. The repeats in the central region are necessary for the interaction of HRC with triadin [258]. The histidine rich region contains two different types of repeats, A and B, which are responsible for binding to the KEKE motif in the luminal portion of triadin. Both A and B repeats have a string of histidine residues followed by several acidic amino acids, the number of which differs between types A and B, followed by either single (type A) or multiple (type B) histadine residues [260]. More than 4.5 repeats are required for triadin binding, although the types of repeats do not affect the strength of interaction.

In addition, the interaction between HRC and triadin is calcium sensitive, such that the strongest interaction *in vitro* occurs in the absence of calcium

[258]. Thus, increases in SR calcium content promote triadin dissociation by disrupting the multimeric structure of HRC [261]. Changes in SR calcium content could provide additional regulation of calcium release by shifting interactions of triadin from HRC to other triadin-binding proteins (e.g., RyR1 and calsequestrin). An area of future research will be to determine whether the dual functions of HRC and calsequestrin (calcium and triadin binding) are redundant or if the two proteins exhibit unique roles in regulating SR calcium storage and release. Defining specific functional roles of different SR calcium-binding proteins is further complicated by the identification of junctate, a third calcium-binding protein in the junctional SR (Figure 30E) [250].

Junctate is a 298 amino acid integral SR membrane protein (33 kDa) with structural homology to junctin. Junctate is derived from alternative splicing of the genes, that encode both junctin and aspartyl -hydroxylase. The initial 78 N-terminal amino acids of junctate are identical to junctin, the first 23 of which are located in the cytoplasm. After a single-pass transmembrane domain, the rest of the protein is located in the lumen of the SR. The C-terminal sequence of junctate is identical to the nonenzymatic C-terminus of aspartyl - hydroxylase. Junctate binds approximately 21 mol Ca^{2+}/mol protein with an apparent kD of 217 M, primarily within the negatively charged luminal C-terminus of the protein [250]. Thus, like calsequestrin and HRC, junctate is a high-capacity, moderate-affinity SR calcium-binding protein and therefore contributes to the calcium storage/release capacity of the SR. However, unlike calsequestrin and HRC, the calcium binding sites of junctate are tethered close to sites of calcium release by anchoring of the protein in the junctional SR membrane via the transmembrane domain of junctin. Because mRNA transcripts of junctate are less abundant in skeletal muscle compared with pancreas, heart, brain, and kidney, junctate most likely plays a limited role in SR calcium storage in skeletal muscle [250]. The concentration of the calcium-binding proteins calsequestrin, HRC, and junctate in the junctional SR serves to significantly enhance the level of high-capacity calcium storage close to sites of calcium release. However, this does not preclude an important role for calcium-binding proteins within nonjunctional regions of the SR.

Sarcalumenin (SAR) is a high-capacity (35 mol Ca^{2+}/mol protein at pH 7.5), moderate-affinity (kD of 300 M) calcium-binding protein that is primarily located in the longitudinal SR (Figure 30F) [242]. Sarcalumenin is a splice variant of an incompletely characterized 53- kDa glycoprotein located in the lumen of the longitudinal SR. The full-length amino acid sequence of SAR (436 acidic amino acids) includes a calcium binding insert sandwiched between an N-terminal signal sequence and several putative nucleotide binding motifs for P-loop– containing ATPase/ GTPase in the carboxyl-terminal region of the protein. Sarcalumenin colocalizes with SERCA, suggesting that the protein not only contributes to calcium buffering within the longitudinal SR, but may also influence calcium reuptake. These conclusions are supported by recent results obtained from SAR knockout mice. To date, no mutations in the SR calcium-binding proteins just discussed have been associated with a human skeletal muscle myopathy.

However, in a few families, autosomal-recessive catecholamine-induced cardiac polymorphic ventricular tachycardia (CPVT) is associated with mutations in cardiac calsequestrin. One mutation involves the substitution of a negatively charged aspartic acid residue for a histidine at position 307 [262]. This aspartate residue is located in the negatively charged C-terminus of the protein, and mutation of this residue to histidine may interfere with calcium binding. It is proposed that disruption of calcium binding to cardiac calsequestrin by this mutation may lead to SR calcium overload, non evoked oscillatory SR calcium release, and subsequent triggering of arrhythmogenic early and delayed after depolarizations [262]. In addition, CPVT in three other families has been linked to three different nonsense mutations that each lead to the introduction of premature stop codons in calsequestrin. Given the critical role of SR calcium-binding proteins in controlling SR calcium content and regulating SR calcium release in skeletal muscle, it would not be surprising if future work were to identify mutations in calsequestrin, HRC, junctate, or sarcalumenin, that contribute to calcium dysregulation and contractile dysfunction in certain forms of skeletal muscle disease.

Ryanodine receptors (RyR1s or calcium release channels) located in the terminal SR provide the primary means of SR calcium release during skeletal muscle EC coupling. Although calcium release through IP3 receptors contributes to long-lasting elevations in nucleoplasmic calcium, that regulate calcium-dependent gene transcription.The primary role of RyR1 activity/ regulation is controlling calcium release during EC coupling. The increase in myoplasmic calcium during a single twitch results almost exclusively from calcium released from RyR1 channels located in triads, regions of close apposition between terminal portions of the SR and both sides of a dissecting transverse tubule (T-tubule). Within these junctional regions, alternate RyR1 release channels are associated with a tetrad of dihydropyridine receptors (four DHPRs or L-type calcium channels) in the T-tubule membrane. The triads are located at the A–I band junction and represent sites of EC coupling in skeletal muscle. Skeletal muscle e-c coupling is controlled by a unique, bidirectional interaction between T-tubular DHPRs and RyR1s, located in the terminal cisternae of the SR. When an action potential spreads down the T-tubule, membrane depolarization causes voltage-driven conformational changes in the DHPR, that trigger activation and opening of nearby SR calcium-release channels. Calcium is then released from the SR *via* a process termed voltage-gated calcium release (VGCR), and the increase in intracellular calcium initiates cross-bridge cycling and muscle contraction. Relaxation occurs when calcium is pumped back into the SR by sarco(endo)plasmic reticulum Ca^{2+}-ATPase (SERCA) pumps. The DHPR in skeletal muscle functions both as a sarcolemmal L-type calcium channel and as a voltage sensor, that initiates calcium release through RyR1 during EC coupling. In a sense, the DHPR can be considered the ultimate regulator of calcium release during EC coupling.

Through stimulation of the DHPR, an action potential (AP) initiates full activation of RyR1 for constant, maximal calcium release. However, stimuli subsequent to the initial AP release less calcium than the first AP due to calcium inactivation of calcium release. Release is reduced or terminated before stores are fully depleted to keep constant the amount of calcium released. The coupling of DHPR activation to RyR1 channel opening is thought to be mediated by a physical interaction between the intracellular II–III loop of the DHPR 1S-subunit and as yet undefined regions of the cytosolic aspect of RyR1. However, critical contributions of other regions of the DHPR 1S-subunit and the C-terminal region of the DHPR 1a-subunit to RyR1 coupling have also been suggested. In addition, a potential interaction between calmodulin-binding domains located in both RyR1 and the C-terminal region of the DHPR 1S-subunit has been reported. RyRs were first identified based on their high affinity binding to the plant alkaloid ryanodine. Electrophysiological characterization of RyRs reSarcopla veiled, that these channels are poorly selective cation channels. They possess weak selectivity among divalent cations or among monovalent cations, and only modest selectivity for divalent over monovalent cations. Calcium competes with Mg^{2+}, Na, and K for pore occupancy. However, promiscuity in ion selectivity enables the channel to exhibit a very high unitary conductance (500 pS in symmetrical solutions with monovalent charge carriers and 100 pS in asymmetrical solutions with divalent charge carriers). This high conductance permits rapid and substantial calcium release from the terminal SR during e-c coupling.

Three RyR isoforms with distinct tissue distributions have been identified: RyR1 is primarily expressed in skeletal muscle; RyR2 is the major isoform found in cardiac muscle; and RyR3 is found in small amounts in a wide variety of tissues. For purposes of this chapter, focus is placed on the skeletal muscle RyR isoform [486], RyR1 (Figure 30G). The skeletal muscle SR calcium-release channel is a homotetramer of RyR1 protomers. The human RyR1 protomer contains 5,038 amino acids and has a molecular weight of 565 kDa. The C-terminal one-fifth of each RyR1 protomer contains the transmembrane segments that anchor the protein in the terminal SR. The most recent topological model of RyR1 predicts eight transmembrane helices. The N-terminal approximately four-fifths of the protein is located in the space between the SR and T-tubule and interacts with the DHPR and a variety of other junctional regulatory proteins. RyR1 channel activity is regulated by both cytosolic and luminal calcium. Calcium release through RyR1 exhibits a bell-shaped dependence on cytosolic calcium; low calcium (1–10 M) activates calcium release, whereas high calcium (1–10 mM) inhibits release. However, because the myoplasmic calcium concentration rarely ever reaches millimolar levels, the effect of high calcium on RyR1 activity is most likely manifested under physiological conditions as an inhibition of release by cytoplasmic magnesium binding to a nonspecific low-affinity site. In contrast, luminal SR calcium regulates channel activity through two mechanisms: (1) a feed-through mechanism, in which calcium released from the SR binds to the activation and inactivation sites on the cytoplasmic face of the channel; and (2) a luminal calcium sensor that increases RyR1 channel activity, when SR calcium reaches millimolar levels. The precise mechanism by which RyR1 activity is regulated by luminal calcium is unknown, but appears to involve both direct effects of calcium on the channel and calcium-mediated conformational coupling through a complex quaternary interaction between calsequestrin, junctin, triadin, and RyR1. Recent results indicate that calsequestrin acts as an RyR1 luminal calcium sensor by inhibiting release channel activity at normal SR calcium levels (1 mM), and that this inhibition is relieved at high (4 mM) SR calcium levels. The influence of calsequestrin on RyR activity appears to be transduced through a complex conformational coupling mediated by junctin and triadin. Triadin and junctin bind the RyR in a calcium-independent manner and prime the channel for activation. At physiological levels of SR calcium, the binding of calsequestrin to the triadin–junctin–RyR complex inhibits calcium release. However, interactions of calsequestrin with triadin and junctin are calcium-sensitive, such that extreme low (10 M) and high (4 mM) levels of luminal calcium promote a conformational change in the protein that weakens its interaction with triadin and junctin, thus relieving the inhibitory influence of the quaternary complex on calcium release. However, even small alterations in luminal SR calcium levels during release may result in subtle conformational changes in calsequestrin that are sufficient to influence calcium release during e-c coupling. Effects of SR calcium load on calcium release are also important during normal activation of release by the voltage sensor. In fact, DHPRs can activate RyRs over a broad range of SR calcium levels and thus fully deplete SR calcium, resulting in a dramatic change in calcium release during e-c coupling. The ability of the voltage sensor to fully deplete the SR is likely mediated by long-range allosteric effects, by which luminal calcium levels alter inhibition of release-channel activity by cytosolic magnesium. Although calcium release during e-c coupling is controlled by the DHPR, a number of soluble molecules (e.g., Mg^{2+}, ATP, Ca^{2+}, and reactive oxygen species) and associated proteins (e.g., FKBP12, calmodulin, homer, and PKA) also regulate the activity of the SR calcium-release channel. Cytoplasmic ATP activates, whereas cytoplasmic Mg^{2+} inhibits channel activity. Both the long and short form of the homer protein activate calcium release through RyR1. Homer multimerizes at its C-terminal coiled– coil and leucine zipper domains to form a junctional signaling complex with RyR1. FK 506 – binding protein (FKBP12) binds stoichiometrically to each protomer of the RyR1 tetramer. The binding of FKBP12 to RyR1 profoundly influences SR calcium release-channel function by stabilization of the fully open and closed states of the channel, controlling channel gating frequency, and modulating sensitivity to activation by Ca^{2+}. Together, the effects of FKBP12 on RyR1 function enhance the gain of skeletal muscle e-c coupling. Another important RyR1-regulatory protein is calmodulin, which also binds stoichiometrically to each subunit of the tetramer. Calmodulin activates the skeletal muscle calcium-release channel at low (nanomolar) concentrations of calcium and inhibits the channel at high calcium (millimolar) concentrations. However, the precise role of the calmodulin interaction with RyR1 in modulating calcium release during EC coupling is still unclear. Although specific mechanisms remain to be clarified, RyR1 activity is also regulated by a variety of different post-translational modifications including oxidation and phosphorylation. RyR1 possesses a number of highly reactive sulfhydryl moieties that are susceptible to oxidation and reduction. The combined response of these hyper reactive thiols empowers the RyR1 tetramer to act as an SR redox sensor, resulting in regulation of release-channel activity in isolated systems by changes in the SR transmembrane redox potential. However, redox modulation of SR calcium release during e-c coupling in intact muscle might be of little consequence, because changes in the SR transmembrane redox potential do not directly alter the amount of calcium released from the SR during normal e-c coupling. RyR1 is also known to be phosphorylated by protein kinase A (PKA) and calmodulin kinase II (CaMKII), the effects of which are still highly controversial. PKA phosphorylation has been found to both activate, inactivate, or produce no effect on RyR1 activity. In addition, PKA phosphorylation at Ser2843 has been suggested to trigger FKBP12 dissociation from the release channel and subsequent loss of FKBP12 regulation. However, other investigators have found that PKA phosphorylation at Ser2843 does not alter RyR1 function or promote FKBP12 dissociation. Similarly, contradictory effects of CaMKII phosphorylation on RyR1 channel activity have also been reported. SR calcium release-channel activity in skeletal muscle is modulated by the direct action of a number of distinct pharmacological agents, including ryanodine, caffeine, 4-chloro-m-cresol, and ruthenium red.

Ryanodine binds with high affinity to the C-terminal pore-forming region of the channel. Ryanodine exerts a complex, dose-dependent effect on the activity of the release channel. At low concentrations (10 nM), ryanodine increases the frequency of single-channel openings in the absence of an effect on unitary conductance. At intermediate concentrations (1 M), ryanodine locks the channel in a long-lived open state with a decreased unitary conductance. High doses of ryanodine (100 M) abolish ion conduction through the channel.

Although not a particularly effective calcium-release trigger, caffeine is a low-affinity RyR agonist that stimulates calcium release from the SR by binding to the release channel. Binding of caffeine to the release channel induces conformational changes in RyR1 that result in increased release-channel sensitivity to activation by other activators such as calcium and voltage.

4-Chlorom-cresol, an agent originally used as a preservative in commercial preparations of certain intravenous drugs, activates calcium release through RyR1 and RyR2, but not RyR3, through interactions with the C-terminus of the channel.

In contrast, ruthenium red blocks SR calcium release by binding to the ion-conduction pore of the channel. Given its highly cationic nature, it is possible that, in addition to its effect of blocking the channel pore, ruthenium red may also alter channel regulation through interactions with calcium- and magnesium-regulatory sites.

In nonexcitable cells, depletion of intracellular calcium stores activates a store-operated calcium entry (SOCE) pathway that is used to replenish intracellular calcium stores. SOCE is also observed in excitable cells including skeletal muscle. Although calcium elevations during single-twitch contractions are due exclusively to voltage-gated SR calcium release, extracellular calcium influx through SOCE may be important during prolonged tetanic stimulation and fatigue. SOCE in skeletal muscle is regulated by RyR1 as it is markedly reduced in myotubes derived from RyR1/RyR3 double-knockout mice. Junctate may also modulate store-operated calcium entry by forming a complex with the IP3 receptor and the transient receptor potential protein (TRPC3). In addition, SOCE in skeletal muscle requires normal triadic integrity because it is attenuated in myotubes in which sarcolemma–SR junctions are disrupted. A second, novel calcium entry pathway activated by membrane depolarization, that requires conformational DHPR-RyR1 coupling was recently identified in skeletal muscle (excitation coupled calcium entry, or ECCE). Although ECCE channels exhibit a pharmacology that is strikingly similar to that of SOCE (blocked by Gd3, SKF-96365 and 2-APB), unlike conventional SOCEs, ECCE is activated by prolonged or repetitive depolarization, but not store depletion. The dependence of both SOCE and ECCE on RyR1 activity raises the intriguing possibility that a conformational signaling interaction between multiple triadic proteins simultaneously controls both calcium release, during e-c coupling and activation of both capacitative and noncapacitative calcium entry pathways. Because the triad junction is the primary site of muscle e-c coupling, formation and stabilization of the triad structure is essential for proper coupling between the DHPR and RyR1. Establishing contact between the sarcolemma and SR and subsequent organization of these membranes into triads is thought to be facilitated by a group of integral SR membrane proteins called junctophilins.

Junctophilins have two major domains: (1) an N-terminal cytoplasmic domain, which constitutes the bulk of the protein; and (2) a C-terminal hydrophobic, single-pass transmembrane domain. The cytoplasmic domain of junctophilin contains several novel protein-interaction elements called MORN (membrane-occupation-and-recognition-nexus) motifs, which have the sequence Tyr-Gln/Glu-Gly-Glu/Gln-Trp-x-Asn-Gly-Lys-x-His-Gly-Tyr-Gly. Junctophilin MORN motifs are thought to be responsible for binding to the T-tubular membrane, possibly through interactions with specific membrane proteins or phospholipids [248]. Four separate junctophilin subtypes (JP-1, JP-2, JP-3, and JP-4) have been identified. JP-1 and JP-2 are both expressed in skeletal muscle, although the function of JP-1 has been investigated in greater detail (Figure 30H). JP-1 is specifically implicated in the formation of triad junctions in skeletal muscle because electron micrographs of mouse skeletal muscle lacking JP-1 show abnormal SR structure as well as a reduction in the number of triad junctions. Deficiency in JP-1 contributes to disruption in the efficiency of signal conversion during EC coupling, and thus results in reduced force generation during muscle contraction.

JP-45 is a novel integral membrane protein of the SR junctional face membrane in skeletal, but not cardiac, muscle (Figure 30I). JP-45 also has a single transmembrane domain, a 122 amino acid cytoplasmic N-terminus, and a short luminal C-terminus. JP-45 is phosphorylated by PKA and interacts with both calsequestrin and the C-terminus of the DHPR 1S-subunit (Figure 31), suggesting that this protein may regulate calcium release during e-c coupling and store-operated calcium entry during *tetanus* and fatigue.

Clearance of calcium in skeletal muscle following release during e-c coupling primarily involves calcium reuptake into the SR by the sarco(endo)plasmic reticulum Ca^{2+}-ATPase (SERCA). Although calcium extrusion *via* plasma membrane Ca^{2+}-ATPase (PMCA) pumps and Na^{+}/Ca^{2+} exchangers balances calcium influx during e-c coupling in cardiac muscle, these calcium transport processes play only a minor role in the clearance of myoplasmic calcium in skeletal muscle. SERCA-mediated SR calcium reuptake controls the rate of muscle relaxation in skeletal muscle and is responsible for maintaining a 10,000-fold calcium gradient across the SR membrane. SERCA is a member of the family of P-type Ca^{2+}-ATPases, named so because these calcium transport proteins undergo autophosphorylation in which the phosphate of ATP is transferred to a highly conserved aspartyl residue in the cytoplasmic portion of the protein. Thus, phosphorylation is an essential step in the mechanism of calcium reuptake through SERCA. There are five primary isoforms of SERCA encoded by three separate genes: ATP2A1 (SERCA1), ATP2A2 (SERCA2), and ATP2A3 (SERCA3). SERCA1 is the isoform found in fast-twitch skeletal muscle. The last two exons of ATP2A1 are alternatively spliced to generate both adult (SERCA1a) and neonatal (SERCA1b) skeletal muscle isoforms. Similarly, alternative splicing of the SERCA2 gene also results in the production of two C-terminal splice variants (SERCA2a and SERCA2b) as well as two relatively rare SERCA2 isoforms, SERCA2c70 and SERCA2d. SERCA2a is the principal isoform present in cardiac, slow-twitch skeletal, and neonatal skeletal muscle. SERCA2b is found in both nonmuscle tissues and smooth muscle. SERCA3 is broadly distributed in platelets, lymphoid cells, and some endothelial cells. Despite significant structural and primary sequence similarities, the tissue-specific expression patterns of each isoform suggests, that the different SERCA variants exhibit functionally distinct properties.

The crystal structure of SERCA1a, a 994 aminoacid (110 kDa) monomeric protein located in the membrane of both terminal cisternae and longitudinal SR, has recently been solved. The single subunit contains 10 transmembrane segments, with the bulk of the protein, including the N- and C-termini, located in the cytoplasm (Figure 30J). The structure can be divided into three parts: a large cytoplasmic headpiece connected to an anchored SR transmembrane region *via* a stalk domain. The cytoplasmic head is subdivided into three distinct domains: (1) the actuator (A) domain (residues 1–50 and 131–238); (2) the nucleotide binding (N) domain (residues 360 – 604); and (3) and the phosphorylation (P) domain (residues 330 –359 and 605– 737) (Figure 30J). The actuator domain facilitates engagement of the gating mechanism and thereby regulates calcium binding and release. The phosphate of ATP is transferred to Asp351 in the P domain whereas the adenosine moiety binds to a site in the N domain. The luminal linkers between the transmembrane segments are short, except for the loop between M7 and M8. A number of residues are essential for coordinating the cooperative binding of two closely spaced calcium binding sites (5.7 Å apart) in the transmembrane region of the protein. Site I is formed by Asn768, Glu771, Thr799, Asp800, Glu908, and the oxygen atoms of two water molecules. Site II is formed by residues in M4 and M6 (Glu309, Asn 796, and Asp800) and three main-chain oxygen atoms from M4 [246]. Two calcium ions bind with high affinity to sites I and II on the cytoplasmic side of the protein at the beginning of the reaction cycle and are ultimately released from the luminal side of the protein following ATP hydrolysis. For each ATP molecule hydrolyzed per reaction cycle, two Ca^{2+} ions from the cytoplasm are pumped into the SR lumen in exchange for two or three H^+ ions. In this way, SERCA1a utilizes the energy stored in ATP to clear myoplasmic calcium during muscle relaxation. SERCA1 copurifies with the low-molecularweight regulatory protein, sarcolipin (SLN).

Sarcolipin (Fig. 1K) is a 31 amino-acid protein with sequence homology to the SERCA2-regulatory protein phospholamban (PLN). SLN structure includes a cytoplasmic N-terminus, a single transmembrane domain, and a luminal C-terminus that contains an Arg-Ser-Tyr-Gln-Tyr (RSYQY) ER/SR retention sequence. Similar to the effect of PLN on SERCA2 function, SLN is thought to inhibit SERCA1 activity by decreasing its affinity for calcium and lowering its V_{max}. SERCA1 is also inhibited by NO and its calcium affinity decreases, as pH is reduced. Additionally, SERCA1 function may also be regulated by the luminal calcium-binding protein, sarcalumenin (SAR; Figure 30F). Specifically, SAR interacts with SERCA1, and SR vesicles isolated from skeletal muscle of SAR-deficient mice exhibit decreased calcium accumulation due to a reduction in SERCA protein expression. However, because SERCA mRNA levels in skeletal muscle are similar in wild type and SAR-deficient mice, the interaction of SAR with SERCA may serve to increase protein stability and reduce SERCA degradation. Thus, in the absence of SAR, a reduction in SERCA1 protein expression results in a decrease in SR calcium reuptake and, subsequently, a slower time course for contractile decay. Compared with calcium reuptake into the SR, mitochondrial calcium uptake, plays only a minor role in myoplasmic calcium clearance during muscle relaxation. Nevertheless, mitochondrial calcium levels increase significantly following calcium released from the SR, not only during pathological Ca^{2+} overload, but also following physiological elevations in myoplasmic calcium, that occur during muscle contraction. Because the total calcium-binding capacity of the mitochondria is very small compared to the SR, mitochondrial calcium uptake represents only a very small fraction of total calcium sequestration, that occurs during EC coupling. Nevertheless, this uptake is sufficient to raise free calcium levels in the mitochondrial matrix, enough to activate three mitochondrial dehydrogenases (pyruvate, NAD-isocitrate, and 2-oxoglutarate dehydrogenases), involved in NADH and ATP generation. In addition, an increase in free calcium in the mitochondrial matrix also stimulates the ATP synthetic capacity of the F1F0-ATPase. Although the physiological impact of this "excitation–metabolism coupling" cycle is not yet fully understood, any calcium-mediated stimulation of mitochondrial ATP generation would serve to lessen the ATP deficit, observed during vigorous muscle activity. Mitochondria possess the machinery required to transport calcium into and out of the matrix.

A large electrochemical gradient (180 mV) provides a substantial driving force for mitochondrial calcium entry, that occurs through several postulated mechanisms, including the mitochondrial calcium uniporter, a rapid calcium-uptake mode (RaM) and a mitochondrial RyR. Because of the low affinity of the uniporter for calcium (half-maximal transport at 30 M calcium), the physiological relevance of mitochondrial calcium uptake has been controversial.

Nonetheless, direct measurements of mitochondrial calcium using targeted calcium probes have demonstrated that mitochondria sequester calcium during cytoplasmic calcium oscillations in fibroblasts, endothelial/epithelial cells, cardiac/skeletal muscle cells, neurons, and pancreatic acinar cells. Efficient mitochondrial Ca^{2+} uptake by the low-affinity uniporter is explained by the presence of tight SR–mitochondria colocalization and the formation of functional calcium microdomains in skeletal muscle. The findings just presented, clearly indicate that proper calcium storage, release, and reuptake are essential for normal physiological functioning of muscle excitation, contraction, and relaxation.

This conclusion is made all the more evident by the fact that several human skeletal muscle disorders (malignant hyperthermia, central core disease, and Brody disease) result from dysfunction in the control and coordination of different aspects of these three critical processes.

Analogous to how an engine governor is designed to maintain a specified speed, the SR functions to maintain vastly different resting myoplasmic and SR calcium levels. This control involves a complex and highly regulated balance of calcium storage, release, and reuptake coordinated by the activity of and interactions between a diverse set of SR calcium regulatory proteins (Figure 31).

Under basal conditions, SERCA pump activity and calcium binding to luminal SR proteins (calsequestrin, HRC, junctate, and SAR) set resting myoplasmic calcium levels (50 – 100 nM) well below that required for myofilament activation and also establish a 10,000-fold calcium gradient across the SR membrane. During muscle excitation, the control of low myoplasmic calcium is temporarily lost as membrane depolarization stimulates rapid and dynamic release of calcium stores through activation of RyR1 release channels located in the terminal cisternae of the SR. As local calcium levels decrease within regions of release, changes in luminal calcium act to self-limit further release through a complex quaternary interaction between calsequestrin, triadin, junctin, and RyR1.

In addition, release is further limited by calcium-mediated inhibition of RyR1 activity by high-calcium microdomains formed on the myoplasmic side of the release channel. Release is ultimately terminated by release channel closure following membrane repolarization. Subsequent muscle relaxation and restoration of low resting calcium levels are then achieved primarily through ATP-driven calcium reuptake into the SR via SERCA1 activity. In this way, the SR in skeletal muscle functions as a dynamic calcium governor that maintains low myoplasmic calcium levels during periods of muscle inactivity. Furthermore SERCA1 activity establishes a large calcium gradient across the SR that is used to drive calcium release and muscle contraction during excitation, and provides for efficient SR calcium reuptake and contractile decline during muscle relaxation. Even seemingly minor abnormalities in the calcium storage, release, or reuptake functions of the SR calcium governor can result in a profound impact on calcium cycling and, ultimately, lead to muscle dysfunction and disease.

Sarcoplasmic Reticulum Function in Cardiac Ventricular Myocytes

In cardiomyocites, the SR plays a key role in cardiac excitation–contraction coupling, and its function is inextricably linked to the structure and function of the adjacent t-tubule system, which contains many of the proteins that determine SR function. Ca^{2+} release from the SR is triggered by Ca^{2+} influx across the cell membrane. I_{Ca} flowing through LTCCs is the primary source of trigger Ca^{2+}. T-type Ca^{2+} current and NCX can trigger Ca^{2+} release from the SR, but they normally play a minor role [263, 264]. This may be, because these proteins are poorly expressed, they are located distant from RyRs [265], or because the relatively low rate of Ca^{2+} influx via these routes, constitutes a poor trigger [266].

Immunohistological and functional studies have shown that LTCCs are located predominantly in the t-tubules in close proximity to RyRs [265] at the dyad: the junctional region (couplon), where Ca^{2+} influx triggers SR Ca^{2+} release. Investigation of I_{Ca} within the t-tubules requires good voltage clamp control of the t-tubule membrane. Poor electrical coupling between the t-tubule and surface membranes, or propagated responses along the t-tubule membrane, could result in voltage escape, and thus erroneous current–voltage relations. However, at least two experimental evidences suggest good voltage control of the t-tubule membrane:

- first, voltage clamp pulses result in a single exponential decay of whole cell capacitance current [267], consistent with tight electrical coupling between the surface and t-tubule membranes;
- secondly, time to peak I_{Na} is unaltered at different potentials following detubulation and little affected by differences in series resistance over the range 1–4 MΩ in control and detubulated myocytes [268].

The lack of effect of small changes of series resistance and, importantly, the observation that detubulation did not alter I_{Na} (which suggests that there were no errors arising from the presence of the t-tubules) suggest, that the t-tubules are adequately voltage clamped. These data suggest that the membrane potential is likely to be uniform during all, but the largest and fastest changes of membrane voltage induced either experimentally using voltage clamp, or during physiological activation of membrane currents, suggesting that I_{Ca}

within the t-tubules can be measured accurately. Orchard and Brette (2008) [268] used the decrease in I_{Ca} following detubulation to calculate that 75–80% of I_{Ca} flows across the t-tubule membrane in a rat myocyte. The dimensions of these cells are nealry 100 x 20 x 20 μm. Assuming that the specific capacitance of the surface membrane is 1 μF/cm^2, they have a total surface area of 18 600 μm^2, with 48 200 Ca^{2+} channels at the cell surface and 143 400 in the t-tubules, giving densities of 3.8 and 24.3 Ca^{2+} channels/μm^2, respectively [269]. In mammalian myocytes, there appear to be 4–10 RyR per Ca^{2+} channel at each dyad, with the Ca^{2+} channels situated apparently randomly with respect to RyR within the junction.

Interestingly, these proteins appear to be located predominantly in the transverse elements of the t-tubule system [270], suggesting inhomogeneity, even within the t-tubules. It has been suggested, however, that the size of the junction may differ in different regions of the cell [271]. To investigate whether this difference is between the t-tubule and surface membranes, EM data of junctional areas [272] and measurements of Ca^{2+} channel numbers were used to calculate that a typical junction in a rat ventricular myocyte contains 35 Ca^{2+} channels, at both the cell surface and t-tubules [270]. This is somewhat higher than other reports (10–25 Ca^{2+} channels), although rat ventricular myocytes have more RyRs per junction than other species [271], so that the RyR:LTCC ratio of 7 is similar to other species.

The constancy of the RyR:LTCC ratio [273, 274] suggests that mechanisms exist to determine co-localization of LTCCs and RyRs. Consistent with the suggestion that junctions at the cell surface and t-tubular membranes contain the same number of Ca^{2+} channels, and that the RyR: Ca^{2+} channel ratio is constant, the gain of SR Ca^{2+} release (Ca^{2+} release for a given Ca^{2+} trigger) is not significantly different at the surface and t-tubule membranes [275]. However, since the majority of Ca^{2+} release sites are located at the t-tubules, they appear to be the major site for Ca^{2+} entry and release, and hence for excitation–contraction coupling: Ca^{2+} sparks occur predominantly near the t-tubules [276] and localized Ca^{2+} release (Ca^{2+} spikes) occurs at discrete sites at the Z-line [277]. The importance of the Ca^{2+} release occurring at the t-tubules has been demonstrated in cells lacking t-tubules, in which the electrically stimulated rise of intracellular Ca^{2+} initially occurs close to the cell membrane and then either diffuses (Purkinje and neonatal cells) [278, 279], or propagates via Ca^{2+}-induced Ca^{2+} release (atrial, cultured, and detubulated cells) [280-281] into the cell.

Fabiato showed that the effectiveness of Ca^{2+} as a trigger for SR Ca^{2+} release depends on the size and the rate of change of Ca^{2+} at the RyR. It has since become clear that the large and rapid Ca^{2+} influx occurring during the first 5–10 ms of I_{Ca} is responsible for triggering SR Ca^{2+} release [282]. The latter part of I_{Ca} appears to load the SR with Ca^{2+} that is released in the subsequent contraction [283]. Inactivation of I_{Ca} is significantly slower in detubulated cells (i.e., at the surface membrane) than in control cells, but this difference is abolished by ryanodine or BAPTA. This suggests that Ca^{2+} released from the SR inactivates I_{Ca} to a greater extent in the t-tubules than at the surface membrane.

This does not appear to be due to: (1) differences in local SR Ca^{2+} release, because the gain of Ca^{2+} release is the same at the two sites; (2) Ca^{2+} depletion within the t-tubules or differences in voltage dependent inactivation, because the difference was abolished when Ba^{2+} was used as the charge carrier; (3) smaller Ca^{2+} influx, because inactivation was little affected by ryanodine in detubulated cells, and was the same in control and detubulated cells in the presence of ryanodine, despite differences in the amplitude of I_{Ca}.

These data have three important implications:

First, I_{Ca} in the t-tubules is large, but inactivates rapidly due to marked Ca^{2+} dependent inactivation; this gives it the characteristics required for a Ca^{2+} trigger.

Secondly, that the more slowly inactivating I_{Ca} at the surface membrane may be important in loading the SR with Ca^{2+}; this will be discussed below.

Third, that regulation of I_{Ca} by Ca^{2+} is more marked at the t-tubules; this implies that I_{Ca} is differentially modulated at the surface and t-tubule membranes, and that the t-tubules are a key site for the regulation of trans-sarcolemmal Ca^{2+} flux by Ca^{2+} released from the SR.

Local regulation of I_{Ca} has been suggested previously, although not differential regulation between the t-tubule and surface membranes. The observation of marked regulation by SR Ca^{2+} release means that the t-tubules may play an important role in cellular Ca^{2+} homeostasis: an increase in SR Ca^{2+} release decreases Ca^{2+} influx via I_{Ca} and increases efflux via NCX, [284-286] thus restoring Ca^{2+} balance (autoregulation) [287]. If I_{Ca} in the t-tubules is inactivated to a greater extent by a given SR Ca^{2+} release than I_{Ca} at the surface membrane, and Ca^{2+} release occurs predominantly at the t-tubules, it appears likely that much of this autoregulation occurs at the t-tubules.

NCX appears to have privileged access to Ca^{2+} released from the SR [288], and the majority of NCX function appears to be located in the t-tubules [289], so that the effect of changes in SR Ca^{2+} release on Ca^{2+} extrusion by NCX may also be most marked in the t-tubule membrane. I_{Ca} may be modulated by other factors, in addition to Ca^{2+} in the dyadic cleft. First, since the t-tubules represent a space with restricted diffusion and a high concentration of Ca^{2+} transport proteins, changes of Ca^{2+} may occur within the t-tubule lumen, which would modulate membrane currents; in this case, the extracellular face of the protein would be exposed to a Ca^{2+} concentration different from the bulk extracellular solution, in the same way that the intracellular face is exposed to a concentration in the fuzzy space that is different from that in the bulk intracellular compartment.

When Ba^{2+} is used as the charge carrier, the rate of inactivation of current through the LTCC is the same in the surface and t-tubule membranes, suggesting that depletion does not contribute to inactivation. However, these experiments were performed at a low stimulation rate to allow recovery from inactivation between pulses. Simulation of ion concentration changes within the t-tubule lumen using biophysically realistic computer models of the cardiac ventricular myocyte incorporating a t-tubule system and known distributions of Ca^{2+} flux pathways and diffusion rates suggest that at physiological stimulation rates Ca^{2+} depletion may occur in the t-tubule lumen, and that this decreases I_{Ca} and the amplitude of the Ca^{2+} transient as a result of the decrease in both the trigger and loading functions of I_{Ca} [290]. The rapid inactivation of I_{Ca} in the t-tubules may help limit such depletion. Secondly, stimulation of I_{Ca} by PKA-induced phosphorylation may be different at the surface and t-tubule membranes. Many of the key proteins are located predominantly at the t-tubules and localized stimulation of I_{Ca} by beta-adrenergic stimulation has been reported previously [291]. More recently, we have shown that the beta-adrenergic agonist isoprenaline has a greater stimulatory effect on I_{Ca} in intact than in detubulated myocytes, suggesting that the beta adrenergic system is better coupled to the Ca^{2+} channel in the t-tubules than at the surface membrane [292].

Interestingly, preliminary data suggest that phosphorylation of RyR following application of isoprenaline is also inhibited by loss of the t-tubules, suggesting that local signalling may underlie RyR phosphorylation. Although not surprising in one way, because RyR forms part of a macromolecular complex that includes kinases and phosphatases, it is surprising in another, because in detubulated myocytes the speed of propagation of Ca^{2+} from the cell surface is increased by isoprenaline [293]. This is due in part to an increase in SR Ca^{2+} load, which increases the rate of propagation, and an obvious explanation is that RyR phosphorylation also contributes to the increased propagation velocity. The explanation for this discrepancy remains to be elucidated. In addition to changes of Ca^{2+} influx brought about by direct regulation of I_{Ca}, Na^{+} and K^{+} channels, by altering action potential configuration, may also alter Ca^{2+} fluxes. Recent work has shown that the neuronal Na^{+} current isoform is located predominantly in the t-tubules, whereas the cardiac form is found predominantly at the surface membrane [294], the functional consequence of these isoform distributions is unclear, particularly since the distribution of total (cardiac + neuronal) I_{Na} is uniform. In contrast, K^{+} currents are evenly distributed between the surface and t-tubule membranes, except for I_{SS}, which is concentrated at the t-tubules [295]. The net effect of detubulation is to decrease action potential duration, with no effect on action potential amplitude or resting potential, presumably due to loss of I_{Ca} and I_{NCX}. Although this suggests that the action potential may be longer in the t-tubules than at the surface membrane, the action potential in the intact cell will be uniform throughout the cell membrane— the mean of that generated by the two membranes, because of their tight electrical coupling.

In addition to triggering Ca^{2+} release from the SR, I_{Ca} loads the SR with Ca^{2+} for subsequent release. Measurements of peak I_{Ca} show 75% of I_{Ca} in the t-tubule membrane [269]. However because of slower inactivation of I_{Ca} in the surface membrane, Ca^{2+} entry across the surface membrane is greater than might be expected from measurements of peak I_{Ca} (e.g., during an action potential in a rat ventricular myocyte, 1.4 mM enters across the surface membrane vs. 3.0 mM across the t-tubule membrane) [269]. Normalizing to junctional area shows that Ca^{2+} entry/μm^2 of junction is actually higher in the surface membrane than at the t-tubules as a result of the slower inactivation (e.g., 1.4 nM in the surface membrane vs. 1.1 nM in the t-tubule membrane during a rat action potential). Importantly, the extra Ca^{2+} entry across the surface membrane occurs during the latter part of I_{Ca}, which has been proposed to be important in loading the SR with Ca^{2+}, suggesting that SR Ca^{2+} loading via I_{Ca} may occur predominantly at the surface membrane. This idea is supported by experiments investigating the effect of detubulation on SR Ca^{2+} loading. Following application of caffeine to deplete the SR, the rate of recovery of SR Ca^{2+} load, and hence Ca^{2+} transient amplitude, is slower in detubulated cells than in intact myocytes [293]. However, SR reloading is slowed by a factor of 2, which more closely reflects the loss of Ca^{2+} entry than the decrease in amplitude of I_{Ca}. Interestingly, in detubulated cells during recovery, the Ca^{2+} transient occurred initially only at the cell periphery, but during subsequent stimuli, the velocity and extent of Ca^{2+} propagation into the cell increased [293]. This suggests that propagation via Ca-induced Ca^{2+} release increases as SR Ca^{2+} content increases. Caffeine-induced Ca^{2+} release, and hence SR Ca^{2+} content, appeared to be the same across the width of the cell, suggesting that the increasing propagation was not due to spatial differences in SR Ca^{2+} content [293].

It also suggests that Ca^{2+} entering the cell at the surface membrane does not cause preferential loading of the peripheral SR, but rather that SR Ca^{2+} content increases throughout the cell, consistent with suggestions that intra-SR Ca^{2+} equilibrates rapidly between regions of the SR.

Ca^{2+} entry following detubulation is reduced by 60%, a value close to the fraction of NCX present in the t-tubules (63%). Since Ca^{2+} entry via I_{Ca} is extruded by NCX during a normal Ca^{2+} cycle, this can explain how SR Ca^{2+} load remains constant after detubulation, and may help reduce changes in $[Ca^{2+}]$ in the t-tubule lumen. However, modelling suggests that at high stimulation rates, Ca^{2+} extrusion is slightly less than influx across the t-tubule membrane so that at steady-state lumenal Ca^{2+} concentration is below bulk cytoplasmic $[Ca^{2+}]$ [290]. This may decrease I_{Ca}, but also implies that under these conditions there must be a net unidirectional Ca^{2+} flux through the cell, from the t-tubule to the surface membrane.

The Ca^{2+} load of the SR will depend not only on I_{Ca}, but also on the relative activity of SERCA and NCX, which compete to remove Ca^{2+} from the cytoplasm. An increase in SERCA activity, or a decrease in NCX activity, will increase SR Ca^{2+} content, and hence Ca^{2+} transient amplitude. SERCA is regulated by PLB, which in its dephosphorylated state inhibits SERCA. Phosphorylation of PLB by PKA (at Ser16) or CaMKII (at Thr17) results in detachment of PLB from SERCA, accelerating Ca^{2+} uptake.

Although biochemical data suggest that SERCA and PLB are distributed throughout the SR membrane [296-298], immunohistological studies show SERCA2b and PLB located predominantly at the Z-line, adjacent to the t-tubules [299]. However, detubulation has no effect on the phosphorylation of Ser16 or Thr17 of PLB in response to the beta-adrenergic agonist isoprenaline [292]. This suggests that PLB is being phosphorylated by a 'global' cell signal, rather than localized signalling like I_{Ca}. NCX is also located predominantly in the t-tubule membrane. It may play a role in Ca^{2+} influx, particularly during the early phase of the action potential [300].

However, its major role appears to be as a Ca^{2+} efflux pathway. Its location at the t-tubules may help ensure rapid and synchronous Ca^{2+} removal from the cytoplasm throughout the cell and may help limit Ca^{2+} depletion in the t-tubules as a result of I_{Ca} in the t-tubule membrane. As the major Ca^{2+} extrusion pathway, NCX plays an important role in determining cellular, and hence SR Ca^{2+} content. The regulation of NCX by phosphorylation is controversial. However, changes in membrane potential and in trans sarcolemmal Na^+ and Ca^{2+} gradients will alter its activity, so that a rise in intracellular Ca^{2+} will increase Ca^{2+} efflux, whereas an increase in intracellular Na^+, or depolarization of the cell membrane will decrease Ca^{2+} efflux [301]. NCX may have privileged access to Ca^{2+} released from the SR: measurement of I_{NCX} and bulk cytoplasmic [Ca^{2+}] during SR Ca^{2+} release has shown that I_{NCX} develops and decays more quickly than the accompanying changes in bulk cytoplasmic [Ca^{2+}]. This suggests that NCX may exist in a compartment, that has restricted diffusion with the bulk cytoplasm, but is closely associated with Ca^{2+} release and uptake by the SR. If so, NCX is situated to extrude Ca^{2+} released from the SR, and to compete with Ca^{2+} sequestration in the SR by SERCA. This may reflect the location of the majority of NCX in the t-tubule membrane, adjacent to, but not at the dyad.

The t-tubules may also influence NCX activity because the concentration of (particularly) inward currents in the t-tubule membrane means that they play an important role in determining action potential configuration and hence NCX function. Thus, NCX may access a Ca^{2+} pool that has restricted diffusion with the bulk cytoplasmic compartment and is shared by RyR and SERCA. Since these proteins appear to be located predominantly at the t-tubules, such Ca^{2+} pools may occur predominantly at the t-tubules. PLB may be predominantly similarly located, where it can regulate SERCA close to the t-tubule membrane, and hence RyR.

Although, immunohistological data suggest that the physical co-localization and hence, presumably, the functional interaction of these proteins is less marked than that of I_{Ca} and RyR, it is tempting to suggest a similar, although less tightly coupled, functional domain, predominantly at the t-tubules, where Ca^{2+} removal pathways compete for Ca^{2+} to be removed from the cytoplasm, and thus help determine SR Ca^{2+} content.

New Concepts of Endoplasmic Reticulum Function in the Heart: Programmed to Conserve [302]

The network of membranes called the endoplasmic reticulum (ER) is a well-studied organelle in various cell types [303]. Since its discovery and visualization by George Palade [304], the rough ER, studded with ribosomes, has been shown to be the major site of secreted and membrane protein synthesis [305]. Proteins synthesized in the ER are routed to the Golgi where they are directed to their final destinations [306]. Although secreted and membrane protein synthesis in the ER has been studied extensively in many cell types [307, 308], it remains largely uncharacterized in cardiac myocytes. A network of membranes similar to the ER, called the SR (Figure 32A) has been defined and studied in striated muscle cells, including in cardiac myocytes [487]. The SR surrounds the myofilaments and operates in collaboration with deep invaginations of the sarcolemma, called transverse (t)-tubules (Figure 32B), to regulate the release of calcium from the SR lumen into the cytoplasm, where it regulates myocyte contraction [309, 310].

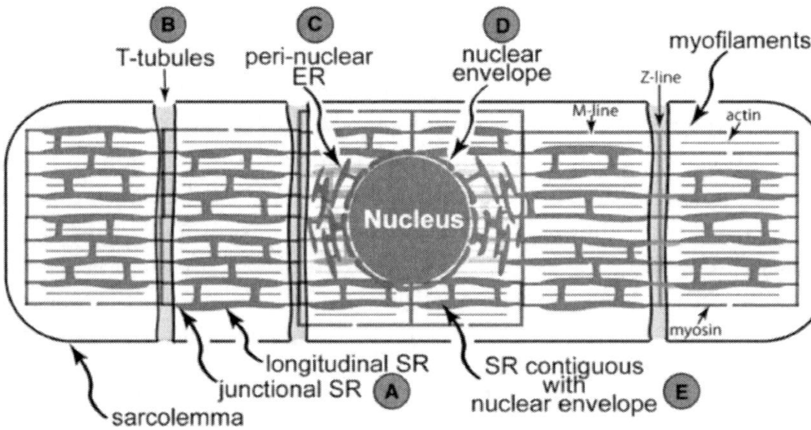

Figure 32. Sarco/endoplasmic reticulum network in a cardiac myocyte: shown is a diagram of a cardiac myocyte depicting the relationships between the region of the SR near the t-tubule, junctional SR, the longitudinal SR (A), transverse, or t-tubules (B), the peri-nuclear ER (C) and nuclear envelope (D), and a depiction of SR that is contiguous with the nuclear envelope (E) The t-tubules are invaginations of the sarcolemma that reside over the Z-line of the sarcomeres [487]. Also shown are the actin and myosin that comprise major portions of myofilaments, as well as the M- and Z-line regions of the sarcomeres. The nuclear envelope and peri-nuclear ER are contiguous, and constitute a location for secreted and membrane protein synthesis, as well as calcium storage and release. The hypothetical localization of secreted and membrane protein synthesis to only the nuclear envelope and peri-nuclear ER, and not to the SR, is shown on the left of the diagram. The hypothetical localization of secreted and membrane protein synthesis to the nuclear envelope, peri-nuclear ER and the SR, depicted as a contiguous membranous system, shaded purple, is shown on the right of the diagram.

However, the relative locations, protein synthetic functions, and protein expression profiles of the ER and the SR in cardiac myocytes are unclear [311, 312]. Some evidence suggests that the ER and, thus, the site for secreted and membrane protein synthesis in cardiac myocytes is in a peri-nuclear network that is contiguous with the nuclear envelop (Figure 32C and D), while other evidence suggests that protein synthesis may also take place in the SR [313–316]. This latter concept is supported by findings that at least part of the SR is physically contiguous with the peri-nuclear ER (Figure 32E), as shown by studies demonstrating that calcium can diffuse freely between the two membrane systems [317]. Secreted and membrane proteins made in the heart have important functions in the heart, as well as in other locations. Since the folding and synthesis of secreted and membrane proteins can be impaired during some cardiac pathologies, there is heightened interest among cardiovascular researchers in the unfolded protein response (UPR), sometimes called the ER stress response [318].

This interest has driven studies aimed at delineating the location of the ER in cardiac myocytes [318]. Electron microscopy has been used extensively to examine the ultrastructure of cardiac myocytes. While this technique may be useful for assessing some aspects of myocyte structure, the location and extent of the ER in cardiac myocytes have been difficult to determine by electron microscopy due to technical limitations [319]. Moreover, while confocal immunocytofluorescent microscopy has thus far played a major role in localizing many SR proteins that are associated with excitation–contraction coupling, it has been used in only a few studies to localize proteins associated with ER protein synthesis, folding, and quality control in myocytes. Such studies have shown that proteins involved in ER protein synthesis, folding, and quality control can be found in peri-nuclear regions of cardiac myocytes, as well as in peripheral areas, where they adopt an SR-like pattern [113–316, 320, 321]. However, additional studies are required to settle fundamental questions about whether the SR serves as a site of protein synthesis, and if not, whether the separate network of membranes around the nucleus in cardiac myocytes is suffi- cient to fulfill the needs of cardiac myocytes to synthesize secreted and membrane proteins. But since it is clear that most secreted and membrane proteins must be synthesized within a network of cellular membranes, the definition of which, in cardiac myocytes, is a topic of ongoing investigation, for the purposes of this review, we will refer to the network in which protein synthesis and folding take place as the ER, and to the unfolded protein response emanating from the ER, as the ER stress response.

The ER Stress Response

Nearly all proteins must be folded into functional configurations [322]. The folding of proteins synthesized in the ER takes place co- and post-translationally, and involves a complex cast of characters that reside in the ER, which constitute the ER protein synthesis, folding, and quality control machinery [323, 324]. An important function of this machinery is sensing the status of ER protein folding by detecting slight changes in the levels of unfolded proteins, and then communicating a status report proactively to the other parts of the cell that respond by adjusting the capacity of the system, thus homeostatically balancing the protein folding demand with the capacity of the protein folding machinery [325]. Drastic changes in the status of protein folding, that threaten this homeostatic mechanism initiate a more reactive version of this response, which is sometimes called the ER stress response [326–330]. The ER stress response can be activated by conditions that alter the ER environment in ways, that impair nascent ER protein glycosylation, disulfide bond formation, or calcium levels; such conditions are often observed in the ischemic, hypertrophic, and failing heart [331, 332] (Figure 33A).

Three ER-transmembrane signaling proteins are major proximal sensors of unfolded proteins in the ER: PERK (protein kinase RNA-like ER kinase) [333, 334], IRE-1 (inositol-requiring protein-1) [335–339], and ATF6 (activating transcription factor 6) [340–342] (Figure 33B–D).

When activated by misfolded proteins in the ER, these sensors facilitate the activation of the transcription factors, ATF4, XBP1, and ATF6 (Figure 33E–G), which mediate the induction of ER stress response genes, that encode ER-targeted chaperones, calcium-binding proteins, and disulfide isomerases, as well as many proteins targeted to other cellular locations. Together, these proteins enhance nascent ER protein folding.

ER stress also suppresses most protein synthesis, while selectively increasing translation of selected mRNAs, most of which encode ER stress response genes (Figure 33H) [343]. This selective translational repression is thought to conserve energy and reduce demands on the ER protein folding machinery [22]. ER stress also augments the ER-associated protein degradation system (ERAD), leading to proteasome-mediated degradation of terminally misfolded ER proteins, which helps relieve ER stress (Figure 33I) [344]. Furthermore, under some conditions ER stress activates autophagy [345–347], an energy-conserving, catabolic process that can promote cell survival [348]. Together, under homeostatic conditions, these aspects of the ER stress response match ER protein folding capacity with demand, which is adaptive (Figure 33J). However, if these aspects of the ER stress response are not sufficient to meet the demand, other ER stress signaling processes guide the cell toward apoptotic cell death, which converts the ER stress response to a maladaptive process (Figure 33K) [349]. Thus, depending on its strength and duration, ER stress can be survival-, or death-oriented.

Interest in the ER stress response in the heart was spawned partly by the realization that myocardial ischemia and cardiac hypertrophy might alter the ER in cardiac myocytes in ways, that would be predicted to impair protein folding in this organelle [350, 351]. Accordingly, studies were carried out to examine the conditions under which ER stress is activated in cardiac myocytes in culture or *in vivo*. For example, ER calcium depletion induces SERCA2 gene expression in cultured neonatal cardiac myocytes [352]. In addition to pharmacological inhibitors of protein glycosylation and compounds that alter ER redox status, ER stress is also activated in cultured cardiac myocytes by maneuvers that mimic pathology, such as simulated ischemia [353].

Moreover, ER stress is activated in border zone cardiac myocytes in the infarcted mouse heart, *in vivo*, in the hearts of mice subjected to pressure overload [351], and in a genetic model of heart failure [354]. Once it was apparent that ER stress was activated in cardiac myocytes under various pathological states, studies were undertaken to examine the functional effects of ER stress in the heart.

For example, several studies have shown that in cultured cardiac myocytes, and in mouse heart, *in vivo*, the ATF6 branch of the ER stress response appears to be adaptive [355]. Transgenic mouse models of ATF6 gain- [355] and loss-of-function [356] have demonstrated that ATF6 is cardioprotective, *in vivo*.

Thrombospondin4 has been shown to be cardioprotective, in part because it is required for ATF6 activation in cardiac myocytes, in vivo [357]. A microarray study showed that in the mouse myocardium, ATF6 induces hundreds of genes encoding numerous ER-targeted chaperones, protein disulfide isomerases, calcium binding proteins, and other proteins, some of which are targeted to the cytosol, as well as changes in the levels of key microRNAs [358].

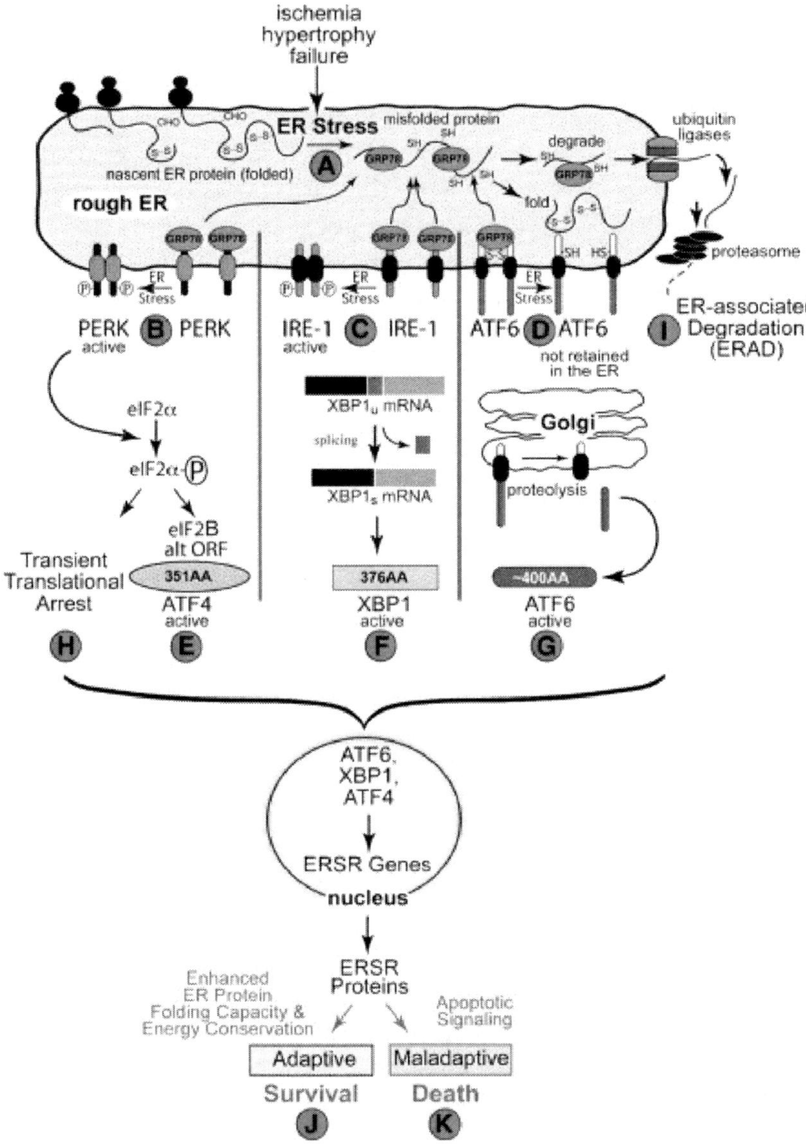

Figure 33. ER stress response signaling: shown is a diagram of the rough ER with attached ribosomes translating mRNAs that encode ER luminal proteins [487]. Conditions that impair the folding of nascent ER proteins, which include ischemia, hypertrophy and heart failure, can result in ER stress (A). Under non-stressed conditions, the ER-luminal chaperone, glucose-regulated protein 78 kDa (GRP78) associates with the luminal domains of the 3 proximal effectors of ER stress, PERK (B), IRE-1 (C) and ATF6 (D). Upon ER stress, GRP78 relocates from the luminal domains of these proteins to misfolded proteins and either facilitates their folding, or escorts them to the degradation machinery. (Continued on next page).

The disassociation of GRP78 from PERK, IRE-1 allows their oligomerization, which fosters trans-phosphorylation and activation of these effectors. In the case of ATF6, dissociation of GRP78 allows ATF6 to relocate to the Golgi, where it is cleaved by site 1 and site 2 proteases which reside in the Golgi. The resulting N-terminal fragment is liberated from the Golgi (G), translocates to the nucleus, and binds to ER stress response elements in ER stress response (ERSR) genes, and regulates their transcription. Activated PERK (C) phosphorylates eIF2α, which fosters transient global translational repression (H) and the translation of the ATF4 mRNA from an alternate start site to generate active ATF4 using an alternate open reading frame (ORF) (E). Activated IRE-1 splices the unspliced form of XBP1 mRNA (XBP1u mRNA) to generate a splice variant form (XBP1s mRNA) which encodes the active transcription factor, XBP1 (F). Like ATF6, XBP1 and ATF6 translocate to the nucleus, and bind to various types of regulator elements in ERSRs to regulate their expression. Depending on the strength and duration of the ER stress, ERSR proteins can foster enhanced ER protein folding capacity, as well as energy conservation, which is sometimes called the adaptive ER stress response, which supports cell survival (J), or ER stress response proteins can lead to cell death, which is sometimes called the maladaptive ER stress response (K).

Although most of these studies support adaptive, protective roles for the ATF6 and XBP1 branches of the ER stress response in the heart, other studies have shown maladaptive effects of ER stress activation in myocardial pathologies (reviewed in a series [359]). For example, β-adrenergic receptor activation has been shown to activate ER stress-mediated apoptosis in cultured cardiac myocytes [360, 361], and δPKC-mediated myocardial damage was shown to be mediated, partly by its effects on activating ER stress [362]. Pressure overload is thought to activate ER stress-mediated apoptosis in the mouse myocardium, and ER stress was shown to contribute to ischemia-induced apoptosis in cultured cardiac myocytes [363].

In addition to ER stress-mediated activation of the canonical unfolded protein response, the ER, which accounts for more than 50% of cellular membrane, serves as a focal point of signaling processes, many of which are oriented toward regulating metabolic signaling and cellular growth.

While there exists an extensive history of studies devoted to examining protein synthesis and quality control in the ER, as well as roles for the ER in regulated calcium release, only recently has it become apparent that the ER plays a regulatory role in cellular metabolism [364]. Indeed, under conditions of increased ER protein and lipid synthesis, which are ATP-utilizing processes, it is reasonable to assume that metabolic pathways responsible for ATP synthesis must be sufficient to meet the increased energy demands.

On the other hand, under conditions of decreased protein and lipid synthesis, such as those resulting from the ER stress response, changes in metabolic pathways decrease energy expenditure accordingly. For example, activation of ATF6 specifically in the heart downregulated 35 of 45 genes in the "fatty acid" and "glucose metabolism" Gene Ontology categories, suggesting that ATF6 activation affects the transcriptome in ways that modulate metabolism.

In part, ER-mediated regulation of energy metabolism is the result of a direct interaction between mitochondria and the ER. There exists an intricate multi-organelle signaling process that involves calcium transfer between the ER and mitochondria [365–368], which is facilitated by a physical association of the two organelles in a structure known as the mitochondria-associated ER membrane, or MAM [369].

Only a small portion of the outer mitochondrial membrane is ER-associated [370], suggesting that a relatively small proportion of ER-derived calcium is transferred to mitochondria. Nevertheless, this direct calcium transfer serves as a mechanism by which mitochondria can sense and respond to conditions in the ER that require adjustments in metabolism [371]. Calcium is released from the ER through the ryanodine and IP3 receptors (RyR; IP3R) [372]. A portion of ER-derived calcium enters a microdomain of the cytosol that is in direct juxtaposition with mitochondria. This positioning, which is the result of physical tethering of the two organelles, involves the binding of the cytosolic chaperone, GRP75, to the ER-associated IP3 receptor and mitochondrial-associated voltagedependent ion selective channel (VDAC) [371].

Studies using non-cardiac myocyte cells established that the ER/mitochondrial tethering also involves the mitochondrial and ER transmembrane GTPase, mitofusin 2 [373]. Thus, calcium released from the RyR or IP3R enters the ER/mitochondrial space, passes through voltage-dependent ion selective channel (VDAC) in the outer mitochondrial membrane, and then through the mitochondrial calcium uniporter (MCU) located in the inner mitochondrial membrane [374]. In the mitochondrial matrix, calcium influences the activities of the Krebs cycle enzymes, α-ketogluarate, pyruvate, and isocitrate dehydrogenases, in ways that increase mitochondrial ATP production [375]. During certain ER stress conditions, calcium release from the ER and, thus, calcium entry into mitochondria is increased, which enhances ATP production and, in so doing, provides a metabolic response that can support adaptive, ATP-requiring aspects of the ER stress response.

A physical connection between the ER and mitochondria has recently been established in cardiac myocytes [376]. The functional consequences of physical interactions of the ER with mitochondria in the heart have been examined in several papers, that employed targeted disruption of the mitofusin 2 gene in mice [377, 378]. Taken together, these studies provide evidence that in cardiac myocytes, calcium transfer from the ER into mitochondria can regulate mitochondrial metabolic function in the mouse heart, *in vivo*. An additional example of how changes in the ER environment could affect metabolic function is the recent finding that mitofusin 2 is an ER stress-inducible protein [379], suggesting that ER stress can enhance the extent of ER/mitochondrial tethering. This is somewhat expected, since increased demands for ER protein synthesis and folding can lead to expansion of the ER and increases in membrane lipid synthesis, both of which are ATP-requiring.

In the heart, one pathological condition that increases ER stress is ischemia. It may therefore affect calcium transport from the ER to mitochondria. Under these conditions, decreased mitochondrial ATP production, impairs ER protein folding, partly by depleting ER calcium. Decreased ER calcium activates proximal sensors of ER stress, because ER chaperones require calcium to fold nascent proteins. A study that investigated the effects of simulated ischemia on endothelial cell injury found that the main pathway of the simulated ischemia induced apoptosis, consisted of the calcium leak from the ER, followed by activation of caspase-12 and caspase-3 [380]. Thus, changes in cellular and ER calcium dynamics have global effects on cellular viability, which could involve signals, that arise from a calcium-depleted ER. Abnormal calcium cycling is known to cause arrhythmia, especially in the setting of transient cardiac ischemia followed by reperfusion. In isolated beating mouse heart, 3 min of ischemia increases diastolic and systolic calcium with a decrease in calcium transient amplitude [381] and increased decay time constant of calcium transient, consistent with inhibition of SR calcium release and re-uptake due to ischemia and acidosis. At the onset of reperfusion, calcium overload is thought to be further exacerbated by uncontrolled influx from the extracellular space [382,383]. A slight but consistent transient increase in diastolic calcium occurs when the amplitude of the calcium transient is decreased. This increase in calcium may have important functional consequences.

Interestingly, Valverde et al. have shown in intact beating hearts that SR calcium release contributes to the transient cytosolic increase of calcium at the onset of reperfusion [381]. This is particularly important in light of interactions between mitochondria and the SR, and could be a potential cause of ER stress during ischemia/reperfusion.

Among the basic principles of protein synthesis is the concept that translation of proteins bound for secretion and membrane proteins occurs on ER-associated ribosomes, while all other proteins are made on free ribosomes. However, puzzling observations from studies carried out using microarray approaches to identify mRNAs, associated with free and ER ribosomes challenged this principle by finding, that some mRNAs that encode cytosolic proteins were found on the ER [384]. A more recent study used cell fractionation and ribosome profiling to sequence all mRNAs associated with cytosolic and ribosomes [385].

In that study, it was found that, in addition to secreted and membrane proteins, ER-associated ribosomes were also engaged in translating mRNAs that encode proteins targeted to other cellular locations. Moreover, it was found that ER-associated ribosomes were more efficient at translating mRNAs. These findings not only expand the scope of protein synthesis on the ER to include the translation of nearly all cytosolic, secreted, and membrane proteins, they redefine the importance of signal transduction systems that monitor the quantity and quality of protein synthesis in the ER, as regulators of the proteome under growth-promoting, as well as growth-limiting conditions.

Under growth-limiting conditions many aspects of the ER stress response program the cell to conserve. As such, the adaptive ER stress response detects potential proteotoxicity due to misfolded proteins, and rebalances the proteome by decreasing overall cellular protein synthesis, and focusing cellular resources on synthesizing essential proteins. Thus, stress signaling from the ER in response to misfolded proteins is predicted to shift metabolism toward maintaining essential cellular functions and modulating cellular growth. Consistent with this prediction are the results of a microarray study that characterized changes in the transcriptome, in mouse hearts upon activation of the ATF6 branch of the ER stress response, which is considered to be adaptive. In terms of growth and metabolism, many growthpromoting genes were downregulated, while genes related to metabolic conservation were upregulated. Thus, one way in which ATF6 is adaptive is that it reprograms cardiac myocytes in ways, designed to conserve energy and limit growth.

Under growth-promoting conditions the mammalian target of rapamycin (mTOR) signaling pathway is a nodal sensor of cellular protein synthesis and, thus, a central regulator of cell growth and metabolism [386]. The effects of mTOR signaling in the myocardium have been well studied and shown to be protective and growth promoting [387–390]. Accordingly, it seems reasonable to assume that cross-talk between ER stress and mTOR signaling might contribute to balancing the proteome with nutritional availability and cellular growth requirements. Generally, mTOR is activated by conditions that are favorable for growth, which makes it an anabolic pathway [386]. On the other hand, ER stress is usually activated by conditions that do not favor growth, which, for the most part, makes it an energy conserving, catabolic pathway. In keeping with their opposing effects on metabolism, although not well established in the heart, many studies of ER stress and mTOR cross-talk in other cell types have shown that the two pathways are often activated in opposition [391].

For example, ER stress is activated by hypoxia or nutrient starvation, while mTOR is inhibited under these conditions [391]. However, the reciprocal activity status of ER stress and mTOR signaling is not always observed, and in some cases, ER stress and mTOR are co-activated [391]. Thus, while there exists cross-talk between ER stress and mTOR signaling, the effect of the cross-talk is apparently conditional, and the molecular mechanisms, as well as the functional consequences of the cross-talk are not well understood. In addition to being regulated by growth-promoting conditions, the magnitude and effects of mTOR signaling are also dependent on the location of mTOR signaling components.

For example, activation of mTOR complex 1, or mTORC1, by the small GTP binding protein, Rheb, occurs on the cytosolic surface of lysosomes, where mTOR signaling is integrated with the availability of amino acids from lysosome-mediated proteolysis [392]. Also, recent studies have shown that mTOR binds to ribosomes [393] and localizes to the ER [393], and that disruption of this localization reduces the anabolic effects of mTOR on protein synthesis [394]. Thus, it is possible that the effects of ER stress and mTOR signaling on each other are not only dependent on the nature of the cellular growth activator or inhibitor, but also on the subcellular localization of signaling complexes. Moreover, it is feasible that the localization of mTOR components to the ER may provide an important spatial dimension to the cross-talk between the ER and mTOR signaling pathways.

Protein synthesis is the major regulator of cell growth, which is a critical parameter of cardiac myocyte function in the healthy and diseased heart. The ER serves as the site for most secreted and membrane protein synthesis, as well as the synthesis of many other proteins, even those that are targeted to other locations. Thus, the ER is a major integration site of cell growth signaling. In cardiac myocytes, the ER membrane network is potentially more expansive than many other cell types, due to the role played by the SR in contractile calcium handling. The potential overlap in function between the SR and the ER in terms of protein synthesis and folding, as well as ER stress and mTOR signal transduction, suggests that the SR and ER membrane system is a macro-organelle, that plays critical roles in cardiac myocyte contraction, growth, and metabolism, all of which are dominant contributors to myocardial function.

Despite its limited role in protein secretion, skeletal muscle contains an extremely extensive network of specialized ER called the sarcoplasmic reticulum [394]. Maintaining the optimal calcium concentration in the sarcoplasmic reticulum lumen is critical to regulating sarcoplasmic reticulum calcium release in the skeletal muscle during contractions. Homeostasis of the ER is not only critical to muscle contraction but also important to maintaining optimal activation of the PKB pathway, which can be inhibited by ER stress. Impairment in PKB activation can lead to anabolic and insulin resistance. With the skeletal muscle being the biggest glucose consuming tissue, it is of utmost importance to keep the ER in minimal stress conditions [394].

Sarcoplasmic Reticulum-Mitochondrial "Through-Space" Coupling in Skeletal Muscle

Muscle contraction requires both Ca^{2+} and ATP. Ca^{2+} ions, which are stored and released by the SR, initiate striated muscle contraction by binding to troponin-C and removing troponin-tropomyosin inhibition of actomyosin interaction. In addition, an increase in mitochondrial Ca^{2+} can also stimulate aerobic ATP production. ATP produced by mitochondria (and glycolysis) is used to drive both crossbridge cycling during muscle contraction and Ca^{2+} reuptake into the SR during relaxation. Thus, regulation of muscle contraction requires careful dynamic control by the SR and mitochondria of sarcomeric Ca^{2+} and ATP availability. Tethering of mitochondria to the CRU provides a physical basis for stable, bidirectional crosstalk between these two organelles (Figure 34).

Ca^{2+} released from the SR stimulates ATP production from the adjacent mitochondrion, which is then used to replenish ATP reserves utilized during muscle activity. Mitochondria, in turn, influence muscle Ca^{2+} signaling by providing ATP required for Ca^{2+} removal, shaping the myoplasmic transient, and inhibiting uncontrolled CICR and local Ca^{2+} release through a modulation of the redox environment of the CRU. Thus, bidirectional SR-mitochondrial communication provides a powerful local control mechanism for integrating Ca^{2+} release/reuptake and ATP utilization during muscle contraction with ATP production and skeletal muscle bioenergetics [396]. Defects in this local control mechanism are likely to result in alterations in proper dynamic and use-dependent control of sarcomeric Ca^{2+} and ATP availability that can contribute to muscle dysfunction during aging, fatigue, and certain forms of muscle disease [397-399].

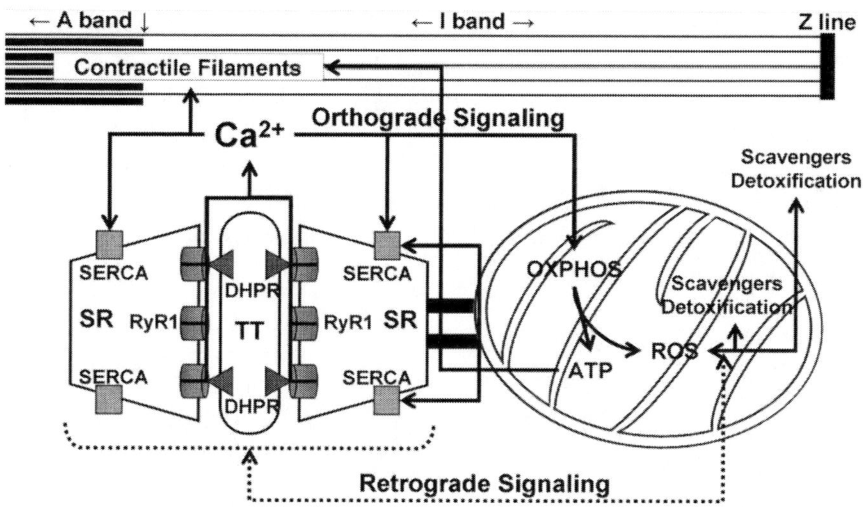

Figure 34. Bidirectional SR-mitochondrial signaling in skeletal muscle [395]. Ca^{2+} release during excitation-contraction (EC) coupling stimulates mitochondrial Ca^{2+} uptake and subsequent ATP production (Orthograde SR-mitochondrial signaling, *solid lines*). Mitochondrial-mediated Ca^{2+} spark suppression (Retrograde mitochondrial-SR signaling, *broken lines*) involves mitochondrial ROS scavenging and detoxification maintaining proper redox balance of the adjacent and tethered CRU. OXPHOS, oxidative phosphorylation; SR, sarcoplasmic reticulum; RyR1, type 1 ryanodine receptor; DHPR, dihydropyridine receptor; TT, transverse tubule; SERCA, sarco(endo)plasmic reticulum Ca^{2+}-ATPase.

Sarcoplasmic Reticulum Function in Smooth Muscle

One of the earliest suggestions that the SR in smooth muscle was the major intracellular Ca^{2+} store and that it acts as in striated muscle was made by Giorgio Gabella and was based on EM studies [400–402]. In the same year, Avril and Andrew Somlyo demonstrated that strontium, used as a surrogate for Ca^{2+}, was accumulated in the SR of smooth muscle (aorta and pulmonary artery), and therefore, it followed that the SR could store Ca^{2+} and be an activator of contraction [403]. Many further important contributions were made by the Somlyos, including the demonstration using electron-probe X-ray microanalysis, that [Ca^{2+}] decreased in the SR following agonist stimulation

[404-406], and that Ca^{2+} cycling to and from the SR occurred during contraction and relaxation [407].

The demonstration of a rich reticular formation in smooth muscle renewed interest in examination of the role of the SR in E-C coupling. Much of this work was looking for or assuming similarity of mechanisms to those of striated muscle. It is only in the last decade that the distinct role of the SR in smooth muscle has been elucidated. Work from Nelson's group [408-410] and others [411, 412] showed that local Ca^{2+} release from the SR can control plasma membrane excitability. We built on this and showed that there is a fundamental relationship between SR Ca^{2+} release and the action potential in ureteric smooth muscle [413].

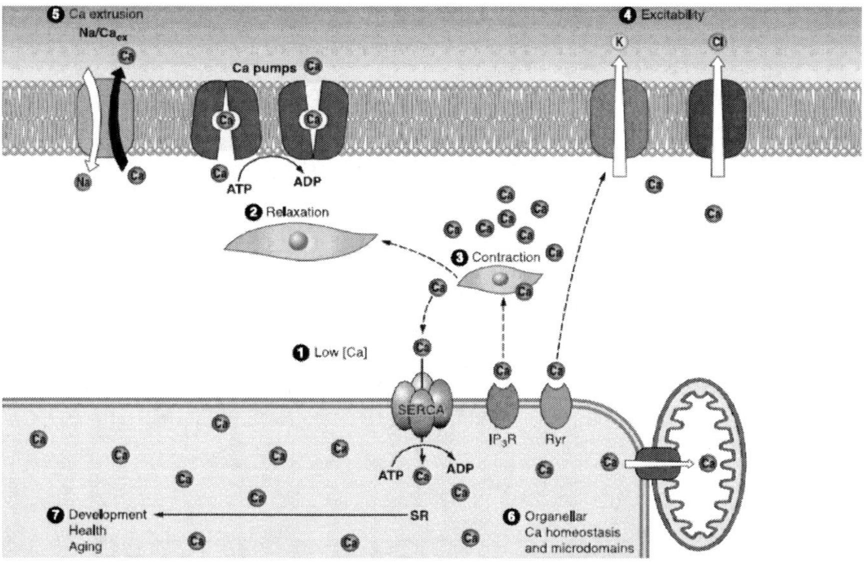

Figure 35. Functions of the SR in smooth muscle. A cartoon to demonstrate the functions of the SR. 1: Contributing to Ca^{2+} homeostasis and maintenance of low resting level of intracellular [Ca^{2+}] via sarco/endoplasmic reticulum Ca-ATPase (SERCA) activity. 2: Contributing to relaxation of the smooth muscle cell by taking up Ca^{2+} via SERCA. 3: Contributing to Ca signals and contraction by Ca^{2+} release via IP3 receptor (IP3R)- or ryanodine receptor (RyR)-gated Ca^{2+} release channels and Ca^{2+} puffs and Ca^{2+} sparks. 4: Contributing to excitability via Ca^{2+}-activated K^+ and Cl^- channels. 5: Facilitation of Ca^{2+} efflux on plasma membrane Ca^{2+}-ATPase (Ca^{2+} pump) or Na^+/ Ca^{2+} exchanger, via vectorial release of Ca^{2+}. 6: Contributing to subcellular microdomains and organellar Ca homeostasis via Ca uptake and release between organelles. 7: Correct SR function needed for normal development and health, and SR functional change is associated with disease and aging.

We now appreciate that the SR can have the following roles [488], as demonstrated in Figure 35:

1) contributing to Ca^{2+} homeostasis by maintaining low resting $[Ca^{2+}]$, *via* action of the SERCA;
2) restoration of $[Ca^{2+}]$ and relaxation following stimulation, via the action of SERCA;
3) augmenting contraction through release of Ca^{2+} and producing or modifying global Ca^{2+} signals in response to agonist stimulation;
4) modulating membrane excitability through activation of Ca^{2+}-sensitive ion channels;
5) contributing to the efficacy of plasma membrane Ca^{2+} extrusion mechanisms by vectorially releasing Ca^{2+} to them;
6) contributing to the maintenance of signaling microdomains around the plasma membrane and other organelles and
7) influencing development, aging, and the health of smooth muscle tissues.

The effects of cell culturing on almost every aspect of smooth muscle E-C are so well known and long standing that investigators ignore them at their peril [414].

Chapter 4

Evolution of the Sarcoplasmic Reticulum

Abstract

The sarcoplasmic reticulum (SR) is crucial for contraction and relaxation of the mammalian cardiomyocyte, but its role in other vertebrate classes is equivocal. Recent evidence suggests differences in SR function across species may have an underlying structural basis. It seems that SR recruitment relates to the structural organization of the cardiomyocyte to provide new insight into the evolution of cardiac design and function in vertebrates.

Sarcoplasmic Reticulum in Vertebrates

The SR is present in all vertebrate classes studied to date including fishes, amphibians, reptiles, birds, and mammals, but the contribution of SR Ca^{2+} release to cardiac contraction varies considerably among species [416]. Briefly, SR Ca^{2+} cycling is more prevalent in atrial vs. ventricular tissue, active vs. sedentary species, adults vs. neonates, and endothermic species (those whose body temperature is dependent on internally generated metabolic heat: mammals and birds) vs. ectothermic species (those whose body temperature is dependent on external sources of heat: fishes, amphibians, and reptiles). The data supporting these general observations has been discussed in detail in earlier reviews [416, 417, 418] and can be appreciated by the work complied by Shiels and Galli [415] in Figure 36 and 37.

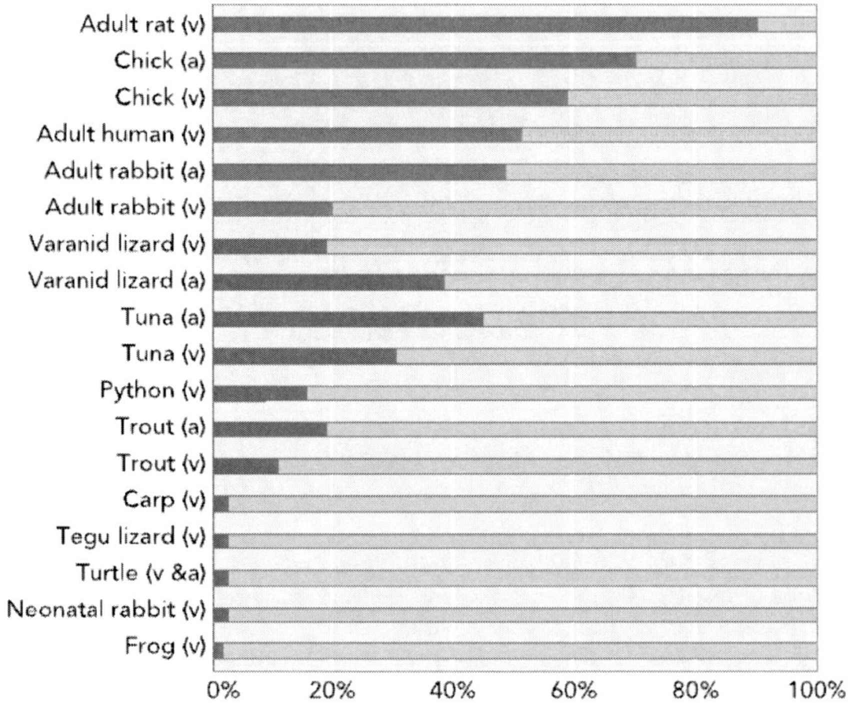

Figure 36. The relative contribution of Ca2+ release [415]. The relative contribution of Ca2+ release from the SR vs. sarcolemmal transport (LTCC and NCX) in the generation of contractile force in isolated cardiac muscle preparations from a number of endothermic (from Adult rat to Adult rabbit) and ectothermic (from Varianird lizard to Frog) vertebrates. These data are compiled from early studies where ryanodine (often in combination with thapsigargin, an inhibitor of SERCA) was used to inhibit force generation in isometric muscle strip preparations. Under these conditions, sarcolemmal Ca2+ entry may overcompensate, thus SR contribution to force may be underestimated for some species. Additionally, stimulation frequency and temperature (both acclimation and experimental) influence the effect of ryanodine on contractile force, which complicates interpretation. Nevertheless, SR dependence across vertebrates correlates well with ultrastructural differences (Table 2 and FIGURE 38), and more recent functional outputs of SR function. v, Ventricular muscle; a, atrial muscle. All preparations were paced at 0.5 Hz, except for human and chick, which were paced at 1.0 Hz. Rat [419, 420], rabbit [421, 422], frog [419], and all reptiles [420] were tested at 30°C. Human [421] and chick [422] were tested at 35–36°C. Trout were tested at 22°C [423], and yellowfin tuna at 25°C [424].

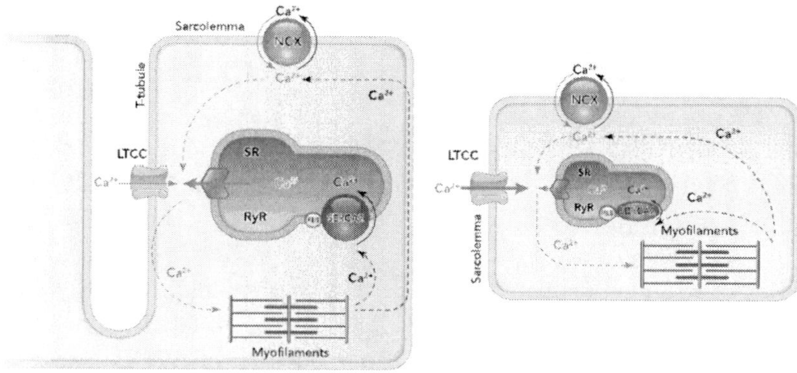

Figure 37. Schematic representation of E-C coupling in endothermic (left) and ectothermic (right) cardiomyocytes [415]. Solid red and blue lines indicate movement of Ca^{2+} during contraction and relaxation, respectively. Dotted red and blue lines indicate cytosolic Ca^{2+} flux during contraction and relaxation, respectively. Briefly, Ca^{2+} is transported across the myocyte membrane (sarcolemma or T-tubules) via Ltype Ca^{2+} channels (LTCC), which releases a larger pool of Ca^{2+} from SR via the RyR. Ca^{2+} can also enter the cell via the Na^{2+}/Ca^{2+} exchanger (NCX). These Ca^{2+} fluxes contribute to the rise in cytosolic Ca^{2+}, which causes myocyte contraction of the myofilaments. Relaxation of the cardiomyocyte occurs when Ca^{2+} dissociates from the myofilaments and is either sequestered into the SR via the SR Ca^{2+} ATPase (SERCA2) or across the membrane via the NCX.

An obvious interpretation is that the SR is critical for the strong and/or fast contractions characteristic of adult endothermic vertebrate hearts, and is not routinely required for the slower, less powerful contractions of most ectothermic vertebrate hearts. All vertebrates have the capacity to store sizeable quantities of Ca^{2+} in their SR, but, paradoxically, ectothermic species, whose contractility is less dependent on SR Ca^{2+} cycling, have greater steady-state and maximal SR Ca^{2+} contents than their mammalian counterparts [425, 426] (see Figure 39).

If all vertebrates can store significant quantities of Ca^{2+} in the SR, why is it released in some species and not others? In other words, what factors limit SR Ca^{2+} release in ectotherms and what is the significance of these stores, if they are not being utilized for E-C coupling? Answers to these questions could provide insight into the evolution of cardiac design and function. Based on the excellent and extensive work of others [427-430], it is suggested that SR spatial organization and SR regulation in ectotherms limits both the magnitude and the rate of Ca^{2+} cycling. Furthermore, correlations exist between SR organization, SR Ca^{2+} cycling, heart rate, and power generation across vertebrates.

The Role of the SR in Mammalian Ventricular E-C Coupling

In the mammalian ventricle, the SR is central to e-c coupling. E-C coupling begins with an AP, which depolarizes the sarcolemmal membrane and drives extracellular Ca^{2+} entry via L-type Ca^{2+} channels [LTCCs or DHPRs]. This small Ca^{2+} entry activates clusters of SR Ca^{2+} release channels (RyRs), causing them to open and release a larger amount of Ca^{2+} from the SR. This process is called Ca^{2+}-induced Ca^{2+} release - CICR and is shown diagrammatically in Figure 36. The close apposition of LTCCs and RyRs is referred to as the couplon [431] and enables local control of Ca^{2+} release from the SR [432]. CICR is assisted by invaginations of the surface sarcolemma, the transverse (t)-tubular system, which transmits excitation deep within the cell (see Table 2).

Table 2. Comparative morphometric data for vertebrate ventricular myocytes [415]

	Lamprey[l]	Trout	Zebrafish[k]	Frog	Turtle[c]	Lizard[d]	Turkey[e]	Rat[a]
Cell length, µm	323	159.8[c]	100	300[f]	189.1	151.2	136	141.9
Cell width, µm	11.9	9.9[c]	4.6	5[f]	7.2	5.9	8.7	32.0
Cell depth, µm		5.7[c]	6.0	n/a	5.4	5.6	n/a	13.3
Capacitance, pF	220	46[b]	26.6	75[g]	42.4	41.2	25.9	289.2
Cell volume, pl	22.6	2.5[b]	2.2	2.9[h]	2.3	2.3	n/a	34.4
SA/V ratio, pF/pl	10	18.2[b]	12	25.8[h]	18.3	18.2	n/a	8.44
T-tubular system	No	No[i]	No	No[g]	No[c]	No[d]	No[j]	Yes[i]

Data are means; SE (when known) has been left out for clarity.
[a]432; [b]433; [c]434; [d]435; [e]436; [f]437; [g]438; [h]derived from 437, assuming an elliptical cross-sectional area; [i]438; [j]439 for finch and humming bird (not turkey); [k]440, [l]441.

In this way, the t-tubular system activates CICR nearly simultaneously throughout the large volume of the mammalian ventricular myocyte and ensures strong spatially and temporally coordinated contractions. Recent imaging in both large (sheep) and small (rat) mammalian ventricular myocytes show adjacent t-tubules are linked via the SR membrane, ensuring synchronicity of ventricular e-c coupling [442]. Relaxation of the mammalian ventricular myocyte occurs when Ca^{2+} is pumped back into the SR via the SR Ca^{2+} ATPase (SERCA) or transported across the sarcolemma by NCX; the sarcolemmal Ca^{2+} pump plays a minor role [443]. Thus, in the mammalian ventricular myocyte, the SR underpins the large and rapid changes in intracellular Ca^{2+} that facilitates the strong and fast heart beat of mammals.

Ultrastructural organization of the sarcolemmal and SR membrane systems underlies both the requirement for, and efficacy of, CICR (Figure 38).

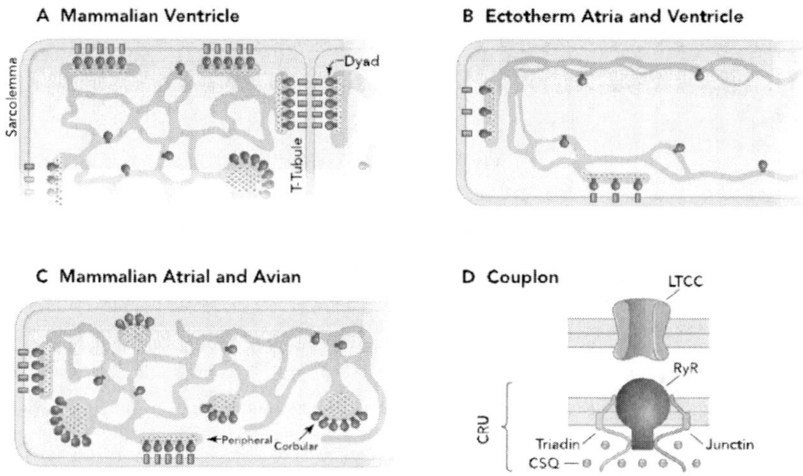

Figure 38. Ultrastructural organization of the SR in a mammalian ventricular myocyte (*A*), an ectotherm atrial and/or ventricular myocyte (*B*), and mammalian atrial and/or avian myocytes (note that atrial myocytes from large mammals contain t-tubules (*C*) [415]. Junctional SR (jSR) contains ryanodine receptors (RyRs), which cluster to form calcium release units (CRUs) that are brought in close proximity to L-type Ca^{2+} CRUs can also exist in corbular or nonjunctional SR (i.e., SR not associated with t-tubules or surface sarcolemma). The CRU is shown as a single RyR for clarity but between 14 and 100 RyRs cluster together to form a CRU and are juxtaposed to many LTCC in the SL membrane (see text for detail). RyRs interact with several proteins, including the Ca^{2+} binding protein calsequestrin (CSQ) and two anchoring proteins, triadin and junctin. Note that differences exist between cell types in the relative proportion of each type of coupling, the distance between couplons, and the overall quantity of SR membrane. Figure created from data in [444] and [445].

Mammalian SR can be grouped into at least three functional and structural elements. Longitudinal/ network or "free" SR (fSR) makes up the majority of the SR membrane and is where Ca^{2+} is pumped into the SR via SERCA. The other two elements are the junctional SR (jSR) and the corbular (cSR) (also referred to as nonjunctional SR, because it is not associated with t-tubules or surface sarcolemma). Both jSR and cSR contain functional clusters of RyRs or "Ca^{2+} release units" (CRUs), defined here as clusters of RyRs. CRUs in jSR are in close proximity to LTCCs, forming couplons at the surface sarcolemmal (peripheral couplings) or along t-tubules (dyadic couplings). Here, the CRUs are activated by Ca^{2+} influx from LTCCs, causing Ca^{2+} sparks, which summate spatially and temporally to produce global Ca^{2+} signals [446, 447].

cSR is present in ventricular myocytes, but it is less prominent than in atrial myocytes. cSR can act as a secondary Ca^{2+} amplification system in response to Ca^{2+} diffusion from jSR release events in both cell types (see Figure 38). Importantly, cSR CRUs can be "silent" until activated, for instance through sympathetic hormones, to coordinate and augment SR Ca^{2+} release throughout the myocyte [448].

Differences in the relative proportion of peripheral, dyadic, and corbular CRUs occur with developmental stage, tissue type, cell size, and adult body size. For example, adult sheep ventricular cardiac myocytes have extensive peripheral couplings, whereas adult rat ventricular myocytes possess extensive dyadic couplings [449]; such differences have been correlated with myocyte size [450, 451] and maximum heart rate [445]. The number of RyRs in mammalian CRUs varies from 14 to 100 tetrameres (a functional RyR is a homeotetramere) [452, 453]. The physical distance between CRUs also varies from between 50 nm [452] and 750 nm [453] (see Table 3).

Table 3. Sarcoplasmic reticulum structural elements in vertebrate hearts [415]

	SR Volume as a % of Cell Volume			CRU			Distance Between CRUs, nm		CRU Alignment	Reference(s)
	jSR	Network SR	Total SR	Peripheral	Dyadic	cSR	jSR	cSR		
Finch v	0.21	0.62	0.83	✓	x	✓	332	126	Z-line	111
Chick v				✓	x	✓	567	235	I-band, A-band	111
Rat v	0.3	3.2	3.5	✓	✓	✓			Z-line	13
Rat a				✓	x	✓	<1,000	<1,000	Z-line	27
Mouse v			6.9	✓	✓	✓			Z-line	13
Mouse a			12.3	✓	x	✓			Z-line	13
Frog v	0.03	0.35	0.38	x	x	x		x	Z-line, M-line, A band	19, 146
Frog a	0.03	0.53	0.56	✓	x	X	>2,000	x	Z-line, M-line, A band	19, 146
Lizard v	0.09	0.6	0.69	✓	x	x	588	x	A-band, random	111
Lizard a	0.11	1.12	1.22	✓	x	x		x		19
Perch v 20°C			4.50	✓	x	x		x		21
Perch v 5°C			6.05	✓	x	x		x		21
Bluefin v 24°C			2.6	✓	x	x		x	Z-line, M-line	123
Bluefin v 14°C			3.7	✓	x	x		x	Z-line, M-line	123
Bluefin a 24°C			3.2	✓	x	x		x	Z-line, M-line	38
Rainbow Trout 14°C				✓	x	x		x	M-line	17
Common Carp 20°C				✓	x	x		x	M-line	32

Values are means (SE, when known, has been left out for clarity). Data are compiled from diverse sources often employing different experimental techniques (e.g., EM vs. immunofluorescence), which complicate some comparisons. When temperature is given, it indicates acclimation temperature. CRU, calcium release unit.

Both of these factors influence spatial and temporal Ca^{2+} signals and the prevalence of propagative CICR between CRUs. Because the distance between

CRUs is small in mammalian ventricular myocytes, activation of CICR at the cell periphery, or throughout the t-tubular system, initiates opening of neighboring CRUs, causing propagated CICR and the possibility of regenerative Ca^{2+} waves [454]. The efficacy of CICR is modulated not only by the structural organization of CRUs, but by their regulation through accessory proteins, numerous small molecules, kinases, and phosphorylation/redox state [455]. Ca^{2+} is the primary ligand, and RyR opening is tightly regulated by cytosolic free Ca^{2+} concentration. Thus RyR Ca^{2+} sensitivity is a critical factor in determining the gain of E-C coupling [456]. The Ca^{2+} sensitivity of RyRs is modulated by PKA, FK506-binding proteins (FKBP12 and FKBP12.6), calmodulin (CaM), CaMKII, and sorcin acting *via* the cytoplasmic domains to alter RyR opening. The effect of PKA phosphorylation on mammalian RyRs can vary, but most suggest increased RyR2 activity.

FKBP12 and FKBP12.6 also affect channel opening; FKBP12 is thought to activate the channel, whereas FKBP12.6 is thought to inhibit the channel [457, 458]. RyR modulation from the SR lumen is achieved via triadin and junctin, which form complexes with calsequestrin.

Calsequestrin is the most abundant Ca^{2+} binding protein in the SR. CSQ2 (the mammalian cardiac isoform) binds Ca^{2+} with a high capacity (60–80 mol Ca^{2+}/mol CSQ2) and a moderate affinity [dissociation constant (kd) of 1 mM] [459], thereby buffering SR Ca^{2+}. In addition to setting SR Ca^{2+} content, CSQ2 directly regulates RyR open probability. Recently, a luminal Ca^{2+}-sensing mechanism, independent of CASQ and dependent on residues in the proposed gate of the receptor, has been reported in mice and controls RyR opening by luminal Ca^{2+}. Although this residue is highly conserved across RyR isoforms, its involvement has not been investigated in ectotherm hearts. The large SR Ca^{2+} content and reluctance of spontaneous SR Ca^{2+} release in ectotherm myocytes (discussed in the next section) may suggest this mechanism, if present, is deferentially regulated.

The Role of the SR in Ectotherm Cardiomyocytes

In the majority of ectothermic species, atrial and ventricular myocyte contractions do not require Ca^{2+} release from the SR. Contraction is supported exclusively by trans-sarcolemmal Ca^{2+} flux through LTCCs, and in some cases with contribution from reverse-mode NCX. Ca^{2+} influx through LTCCs (I_{Ca})

has been estimated at 50 – 80 M cytosol in fish myocytes (expressed per myofibril volume, which is 40–55% of cell volume for carp and trout) and can initiate a full contraction [460]. Sarcolemmal Ca^{2+} entry in mammalian cardiac myocytes is lower (10 M cytosol expressed per non-mitochondrial volume, 65% of cell volume) and alone is insufficient for the activation of normal physiological contractions [460]. Nevertheless, there are ectothermic animals that rely more strongly on SR Ca^{2+} during E-C coupling. For the most part, these are species with elevated cardiac function, or with the phenotypic capacity to elevate cardiac function when required.

For example, varanid lizards are highly athletic reptiles with a functionally divided ventricle capable of developing high systemic blood pressures compared with other squamates. SR Ca^{2+} release in this lizard contributes to both contractile force and the cellular Ca^{2+} transient. Similarly, bluefin tunas are apex oceanic predators with the highest heart rates and cardiac power output among fishes, and this species relies strongly on SR Ca^{2+} cycling during e-c coupling. These generalities in ectotherms correlate with data from endotherms where high activity and metabolism are coupled to high heart rates and SR involvement in e-c coupling (see Figure 37). Thus the importance of SR Ca^{2+} cycling in the hearts of active species appears to be a general principle that can be applied to all vertebrates. SR Ca^{2+} cycling may also be recruited in some stenother mal fish [like burbot, 461], which inhabit cold environments, and in eurythermal fish [like salmonids and tuna, 462], which undergo seasonal cold acclimation (i.e., chronic cold exposure in excess of 1 mol).

Indeed, chronic cooling in both perch and tuna leads to a proliferation of total SR membrane. Increased SR function in the cold ectotherm heart is thought to augment sarcolemmal Ca^{2+} to overcome cold-related reductions in the Ca^{2+} sensitivity of myofilaments and sarcolemmal Ca^{2+} influx. Similar strategies have been described in cold-hibernating mammals [463]. Interestingly, regardless of the relative contribution of the SR to e-c coupling, all ectotherms studied to date possess sizable amounts of Ca^{2+} in the SR (Figure 39).

Several possibilities exist, which may explain the limited SR Ca^{2+} release relative to the high Ca^{2+} storage capacity of ectotherm cardiomyocytes, including SR ultrastructure [density of RyRs, their organization into CRUs, the proximity of CRUs to LTCCs (i.e., formation of couplons)], and SR regulation (RyR sensitivity to opening). The efficacy of trans-sarcolemmal Ca^{2+} flux is aided by the extended length-to-width ratio and high surface area-to-volume ratio of ectotherm myocytes (Table 2).

This elongated or spindle-shaped morphology compensates for the lack of the t-tubular system and is coincident with peripherally located myofilaments. For the majority of sedentary and/or sluggish species of ectotherms, this myocyte arrangement allows for normal contractions to occur without the SR. The lack of a functional role for the SR is also reflected by the low density of RyRs in ectothermic myocytes compared with endotherms. A series of [3H] ryanodine-binding studies by Vornanen and colleagues has shown RyR density is considerably lower in carp (19%), rainbow trout (28%), burbot (65%), and lamprey (68%), when normalized to rat (100%). A similar disparity in RyR density between species has been reported from protein expression studies [464, 465] and broadly corresponds to the relative contribution of SR Ca^{2+} cycling during e-c coupling within each species.

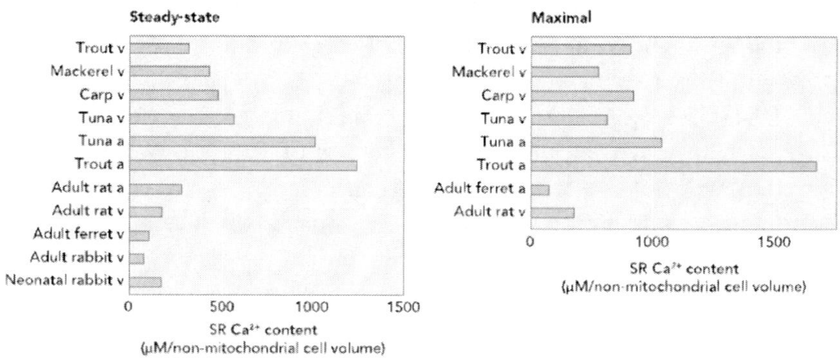

Figure 39. Steady-state and maximal SR Ca^{2+} content in a range of vertebrates SR Ca^{2+} content was measured in voltage-clamped cardiomyocytes and was calculated by integrating the current that flowed during caffeine application [415]. All values are expressed per non-mitochondrial cell volume (which ranges from 40 to 60% of total cell volume). For rat and ferret, original data was normalized to non-mitochondrial cell volume by dividing the reported SR Ca^{2+}content (per total cell volume) by 0.70. Maximum SR Ca^{2+} load for adult rat ventricle and adult ferret ventricle was assessed with application of tetracaine (to inhibit the ryanodine receptor) or with isoproterenol (to achieve maximum Ca^{2+} uptake), respectively. Data are collated from adult ferret *Mustela putorius furo* and adult rat *Rattus rattus* ventricle; adult rat *Rattus rattus* atria; adult rabbit *Oryctolagus cuniculus*, neonatal rabbit *Oryctolagus cuniculus*, bluefin tuna *Thunnus orientalis*, mackerel *Scomber japonicas*, carp *Carassius carassius*, and trout *Oncorhynchus mykiss*.

For example, frog ventricle shows no evidence of SR Ca^{2+} cycling during e-c coupling [466], which is confirmed by the absence of RyRs at peripheral couplings between the sarcolemmal and SR membranes. Similarly, carp demonstrate sparse immunohistochemical staining of RyRs, and functional studies show no effect of SR inhibition on contractility [467]. The cold stenothermic burbot, on the other hand, shows strong SR dependence of contractile force and structural evidence of peripheral couplings [468]. Clearly, the relatively low number of RyRs will be a contributing factor in the weak CICR typical of many ectotherms. At present, the number of RyRs that make up a CRU is not resolved for any ectothermic species. However, the spatial relationships between CRUs and the temporal/spatial properties of cellular Ca^{2+} cycling during e-c coupling have been measured.

EM studies from the green lizard (*Anolis carolinensis*) ventricle show widely separated peripheral CRUs and a lack of cSR [445], but functional work is lacking for this species. EM studies confirmed the presence of peripheral couplings in the bluefin tuna, but the percentage of total SR that forms peripheral couplings was very low (0.98% and 0.40% in the atria and ventricle, respectively), and there was no evidence of cSR. Functional studies in both tuna [469] and trout [470] show CICR occurs at the myocyte periphery but with limited propagated activation of neighboring peripheral CRUs or central CRUs under normal physiological conditions. In other words, the Ca^{2+} wave appears to be carried via diffusion between CRUs along the periphery and into the myocyte. For example, trout ventricular myocytes exhibit a small U-shaped delay (6 ms) in the peak of the Ca^{2+} transient between the periphery and the center under routine conditions.

However, increased Ca^{2+} influx at the periphery after application of BayK8644 (an agonist that increases the open probability of LTCCs) increased the rate and amplitude of centripetal Ca^{2+} transients, suggesting fish myocytes may possess "silent" CRUs, that require inotropic stimuli to activate [470]. Structural evidence is lacking, but support for this idea comes from recent work showing both adrenergic stimulation (500 nM isoprenaline) and low (500 M) levels of caffeine (which sensitize the RyR to cytosolic Ca^{2+} and increases RyR opening) result in propagative CICR, in trout ventricular myocytes and in zebrafish ventricular myocytes [471]. These functional studies strongly suggest that recruitment of CICR can occur in certain ectotherm myocytes. However, the spatial relationships between CRUs and the temporal/spatial properties of cellular Ca^{2+} cycling have not been measured together in the same ectothermic species (see Table 3).

This information would help to clarify the limitations SR ultrastructure impose on ectothermic CICR. Ryanodine receptors (RyRs) are very large (2.2 MDa) homotetrameric assemblies with a large cytoplasmic head (originally described as "feet") that protrudes from the SR membrane into the cytosol, and a transmembrane stalk that forms the channel pore [472].

Tissues can express multiple isoforms; however, RyR2 predominates in mammalian cardiac muscle, and a homologous RyR isoform has been identified in non-mammalian hearts (although the entire amino acid sequence of the non-mammalian cardiac RyR has not been completely determined). The Ca^{2+} sensitivity of RyR opening is an important regulator of CICR. In ectotherm cardiomyocytes, RyR Ca^{2+} sensitivity correlates with the contribution of SR Ca^{2+} to force production. Using [H3]ryanodine binding, Vornanen [473] showed that the Ca^{2+} sensitivity of the rat RyR (kd of 0.16 M) is similar to the cold stenothermic burbot (0.19 M), but considerably greater than that of the lamprey (0.35 M), trout (0.83 M), and carp (1.10 M). The reduced Ca^{2+} sensitivity of RyRs means that a higher systolic Ca^{2+} is required to open the ectothermic RyR, even if the structural organization for CICR exists.

Clearly, the low Ca^{2+} sensitivity of the RyR, coupled with a lower RyR density, is a major factor underlying low levels of CICR in most ectotherm hearts [473]. The low RyR Ca^{2+} sensitivity may also explain the lack of spontaneous SR Ca^{2+} release events (i.e., Ca^{2+} sparks) in ectotherm myocytes. Ca^{2+} sparks are tiny Ca^{2+} signals, which arise from the activation of a CRU, that summate in space and time to form the systolic Ca^{2+} transient. Ca^{2+} sparks are absent in quiescent ectotherm myocytes [trout ventricle; zebrafish ventricle] or in low abundance [trout atrium], despite confirmation of large SR Ca^{2+} stores by rapid application of high levels of caffeine (10 mM, which opens RyRs, causing Ca^{2+} release).

Moreover, strategies known to enhance spark frequency in mammals (6mM extracellular Ca^{2+} or elevated intracellular free Ca^{2+}) fail to elicit Ca^{2+} sparks or Ca^{2+} waves in fish myocytes. However, application of low levels of caffeine (200 μM to 1 mM) increased spark frequency in zebrafish myocytes, whereas higher concentrations (5 mM) synchronized SR Ca^{2+} release. This suggests zebrafish RyRs are organized into CRUs, that exhibit low Ca^{2+} sensitivity with limited propagative coupling, under routine conditions, but can be recruited (i.e., via caffeine) to participate in E-C coupling. The zebrafish RyR also responds to PKA activation with a twofold increase in phosphorylation.

These authors found that RyR phosphorylation increased SR Ca^{2+} release, spark frequency, and Ca^{2+} transient amplitude in half of the zebrafish cardiomyocytes studied. The other half of the myocytes responded to PKA activation via SERCA-dependent changes in SR Ca^{2+} content.

Thus, activation of RyRs after sympathetic stimulation is a mechanism through which ectotherms can recruit SR Ca^{2+}cycling. Indeed, functional studies that show CICR during adrenergic stimulation in trout heart were performed previously and it seems highly probable that adrenergic activation of I_{Ca} and RyRs converge to coordinate and amplify Ca^{2+} release.

Modulators known to affect RyR opening in mammals may increase the efficacy of CICR in ectotherms. FKB12 expression was recently investigated in fish heart [rainbow trout, crucian carp, and burbot] and found to be greater in atrial compared with ventricular tissue, in all three species studied, and could be enhanced by cold acclimation in trout. If FKBP12 sensitizes fish cardiac RyRs to cytosolic Ca^{2+}, then this finding may correlate with the greater SR involvement in e-c coupling apparent in atrial *vs.* ventricular tissue and after chronic cold exposure.

Similar studies are lacking for other non-mammalian vertebrates; however, the association of FKBPs with cardiac RyRs is evolutionally conserved, and both isoforms are found in most species (human, pig, rat, mouse, chicken, frog, and fish), but abundance varies greatly. Regulation may also occur on the luminal side of the SR. As discussed earlier, in mammalian cardiomyocytes, RyR opening is regulated by luminal Ca^{2+} *via* specific residues in the channel pore, that provide a Ca^{2+}-sensing mechanism. This is in addition to the regulation of RyR opening that is regulated by CSQ2.

CSQ has been identified in the SR of fish, amphibians, reptiles, aves, and mammals [445]. It has been cloned and sequenced in a range of vertebrates and appears to be relatively well conserved, suggesting a fundamental biological role. It is possible that CSQ inhibits RyRs to a greater extent in ectothermic myocytes and that a higher concentration of luminal Ca^{2+} is necessary to relieve the inhibition. Although this possibility has not been directly measured, Korajoki and Vornanen [474] have examined the expression of the ectothermic isoform of CSQ in warm and cold acclimated rainbow trout atrial and ventricular tissue. Based on known differences in SR Ca^{2+} function, Korajoki and Vornanen [474] hypothesized that expression would be elevated in the *atrium* compared with the ventricle and would be increased after cold acclimation in both chambers.

However, no differences were detected, suggesting CSQ is not involved in chamber or acclimation specific enhanced SR function in trout. In agreement with this result, mammalian atrial muscle, which is known to be more SR dependent, does not express higher CSQ2 compared with ventricular muscle and CSQ2 expression in the heart remains unchanged in mammalian cardiac pathologies, which are known to alter RyR function [474].

Although these studies indicate that the amount of CSQ does not correlate to SR function in vertebrates, the possibility of differential regulation (e.g., through phosphorylation), differential interaction with RyRs or other luminal proteins. Furthermore, differential CSQ isoforms expression [474] cannot be excluded. Indeed, altered CSQ isoforms expression has been documented in cold-hibernating ground squirrels. Collectively, the data from the limited number of studied ectotherms suggest two major E-C coupling schema:

1) that typified by frog and carp, where there is no (or very limited) CICR during contraction and
2) that typified by ectotherms with enhanced cardiac function like salmonids, where CICR occurs at CRUs in the cell periphery followed by Ca^{2+} diffusion through the cytosol, without the requirement for propagative CICR.

This peripheral initiation and centripetal diffusion appears to be the basal arrangement for cardiac E-C coupling among vertebrates [450] and is sufficient to drive the relatively slow, low-pressure hearts of most ectothermic species. Mobilizing silent CRUs through adrenergic stimulation, chronic cooling, or pharmacological intervention provides a mechanism to increase the rate and strength of ectotherm cardiac contractions in species with elevated cardiac function.

The Mechanism and Significance of the Large SR Ca^{2+} Stores in Ectotherms

Because of the limited capacity to release SR Ca^{2+}, one may expect that ectotherm SR would possess a modest Ca^{2+} content. However, quite to the contrary, the steady-state and maximal SR Ca^{2+} content of ectothermic vertebrates are considerably larger than those reported in mammalian cardiomyocytes (see Figure 39). This difference exists despite the fact that ectothermic SR density per cell volume (ranging between 0.38 and 6.1%) is generally less than that found in mammals (ranging between 3.5 and 12.3%). The smaller SR Ca^{2+} content in mammalian myocytes may be partly explained by leakier RyRs, which limit the maximum amount of Ca^{2+} that can be sequestered at any given time.

Indeed, spontaneous Ca^{2+} sparks are far more common in mammals than in ectotherms. Also, RyRs from fish do not show spontaneous opening in response to cooling; in fact, passive leak from trout SR vesicles decline at cold temperatures. However, when the RyR inhibitor tetracaine is applied to rat ventricular myocytes to abolish Ca^{2+} leak, maximum SR Ca^{2+} content is still at the lower end of those found in ectothermic myocytes (Figure 39). Thus the extent of RyR leakiness cannot fully explain the large SR Ca^{2+} contents found in ectotherms. Other possibilities, such as enhanced SR Ca^{2+} uptake and/or luminal Ca^{2+} buffering, are discussed below.

SR Ca^{2+} uptake is achieved exclusively by SERCA (the cardiac isoform is SERCA2), which consumes ATP to actively pump Ca^{2+} into the SR lumen. The high Ca^{2+} binding efficiency of SERCA2 allows for rapid SR Ca^{2+} uptake and largely determines the rate of mammalian myocyte relaxation. Several isoforms of SERCA2 exist within the mammalian heart (SERCA2a, b, and c), which differ in their enzymatic turnover rate and affinity for Ca^{2+}, but SERCA2a is the dominant cardiac isoform. Few studies have analyzed the total number and sequence of SERCA2-encoding genes in non-mammalian vertebrates. Interestingly, as genome duplication occurred within the fish lineage, teleosts may possess paralogs of the SERCA2 gene, leading to greater diversification of SERCA2 isoforms.

Indeed, melting curves of rainbow trout SERCA2 transcripts reveals three closely clustered peaks, suggesting the trout heart may possess various isoforms of SERCA2, but the functional significance of these isoforms has not been clarified. SERCA2 expression and activity is dependent on numerous factors, including region of the heart, developmental stage, species, environmental conditions, and pathological state. Nevertheless, SERCA2 expression and activity correlates well with the role of SR Ca^{2+} cycling during e-c coupling in vertebrates, showing a similar pattern to that discussed earlier for RyR density and structural organization. In mammals, SERCA protein/mRNA expression and SERCA activity are higher in adult *vs.* neonatal, larger *vs.* smaller animals, and atrial *vs.* ventricular cardiomyocytes. These differences correlate with SR Ca^{2+} content, which is greater in smaller animals and in tissues, that are capable of generating high frequencies of contraction. Mammalian knockout mouse models show reduced SERCA2 expression, which leads to lower SR Ca^{2+} content, whereas overexpression increases it. Thus, if ectothermic animals have a relatively high expression of SERCA2. This may help to explain their large SR Ca^{2+} contents.

Although direct comparison of ectothermic *vs.* endothermic SERCA2 expression has not been made, data regarding SERCA2 activity has shown that this theory is unlikely. In rat and rabbit ventricle, SERCA2 activity is four to six-fold higher at room temperature than a range of active/sedentary teleosts. When corrected for routine body temperature. It has been showen the relative

SR Ca^{2+} uptake rates were 100, 26, 19, 18, 11, and 2% for adult rat, newborn rat, cold-acclimated trout, warm-acclimated trout, warm-acclimated carp, and cold-acclimated carp, respectively. These findings indicate that SR Ca^{2+} uptake is slower in fish compared with mammalian hearts, but that activity can be regulated, in this case by temperature. Indeed, mammalian SERCA2 is regulated by a variety of molecules and hormones, including thyroid hormone, Ca^{2+} concentration, ATP concentration, phospholamban (PLB), sarcolipin, and various pathological states. It is possible that some of these factors may contribute to the superior ectothermic SR Ca^{2+} storage capacity. PLB is a regulatory phosphoprotein that is localized to the longitudinal "free" regions of the cardiac SR. In the dephosphorylated state, PLB acts as an inhibitor (brake) of SERCA2, whereas phosphorylation at Ser-16 and Thr-17 releases this inhibition and promotes Ca^{2+} uptake into the SR by increasing the Ca^{2+} affinity of SERCA2.

Therefore, SR Ca^{2+} content can be regulated by PLB by either altering PLB expression or phosphorylation status. Since PLB is an inhibitor of SERCA2, reduced expression of this protein is expected to increase SR Ca^{2+} content. Indeed, PLB knockout mice have a greater SR Ca^{2+} content, and the ratio of PLB/SERCA2 expression has been correlated with SR function in a number of mammals. Thus, PLB expression may provide a mechanism to enhance SR Ca^{2+} content in the ectothermic heart. The presence of PLB has been confirmed in a wide range of vertebrates, and its molecular structure is well conserved.

For example, the mammalian PLB shares a high sequence homology with chicken (85%), zebrafish (76%), and pufferfish (67%), suggesting that PLB regulation of SERCA2 has early evolutionary origins. Early comparative analysis confirmed chick, frog, and carp PLB content are one to two orders of magnitude lower than that found in the dog, rat, or human heart. However, it is not known whether these differences reflect a mechanism for enhanced SR Ca^{2+} content, or the lower SR density characteristic of these ectothermic species. With respect to phosphorylation status, PLB can be phosphorylated by PKA and/or Ca^{2+}/calmodulin dependent protein kinase II (CaMKII). Phosphorylation most commonly occurs in response to adrenergic stimulation, which directly activates PKA and CAMKII by elevating cAMP and intracellular Ca^{2+}, respectively.

Together, this dual response leads to a potent acceleration of SR Ca^{2+} uptake and forms the Ca^{2+} dependent basis of the adrenergically mediated quickening of cardiac relaxation. An increase in the phosphorylation status of

PLB is therefore another possibility to explain the large SR Ca^{2+} stores of ectotherms.

Recent work with zebrafish ventricular myocytes shows that stimulation with PKA increases SR Ca^{2+} content and leads to greater CICR, indicating that the SR is regulated *via* similar pathways in fish and mammals. Furthermore, Korajoki and Vornanen [475] found higher SERCA2 expression and a lower PLB-to-SERCA2 ratio in the *atrium vs.* the ventricle of the trout heart, which correlates with faster Ca^{2+} uptake in trout atrial muscle.

The basal level of PLB activation in ectotherms has not been determined, and there is no evidence that PLB is phosphorylated to a different extent in ectotherms *vs.* endotherms. Furthermore, adrenergic stimulation has predominantly inotropic effects in the heart of the carp and frog, whereas acceleration of relaxation in these species is either absent or minor compared with the mammalian adrenergic response. If an adrenergically mediated relaxation is observed in the frog heart, the mechanism appears to involve cAMP dependent Na^{+}/ Ca^{2+} exchange and not PLB regulation of SERCA2.

Clearly, further studies are necessary to investigate the correlation between PLB expression/phosphorylation in ectotherm cardiomyocytes.

Ectothermic SR Ca^{2+} Buffering

The mammalian heart contains a range of luminal SR Ca^{2+} buffering molecules including sarcalumenin, calreticulin, calumenin, and CSQ2. Among these, CSQ2 is clearly the most predominant, and recent estimates suggest 75% of Ca^{2+} taken up by the SR is bound to CSQ2. Similar to transgenic studies with PLB, overexpression of CSQ2 in mammalian cardiomyocytes, either acutely or chronically, leads to an increase in SR Ca^{2+} content, whereas a reduction in the expression of CSQ2 by gene silencing results in a decrease in SR Ca^{2+} content. Coupled with the fact that CSQ2 interacts with RyR directly to inhibit Ca^{2+} release (discussed earlier), studies from mammals suggest differential expression or regulation of CSQ is a likely candidate to explain the large ectothermic SR Ca^{2+} contents.

However, as discussed above, the limited studies conducted on ectotherm CSQ have not correlated with high SR Ca^{2+} content. However, these authors calculate the total buffering capacity of trout SR to be 15.8–21.12 mM,which is slightly larger than values from the mammalian heart. Clearly, there is a need to further quantify CSQ content in vertebrates to get a better understanding of species specific-buffering capacities in the SR and relate this to overall Ca^{2+} storage capacity.

Ectothermic vertebrates are frequently challenged with a fluctuating environment that can alter body temperature, oxygen levels, and pH. Variation in these factors in mammalian vertebrates is known to drastically alter myocyte ionic homeostasis. In particular, a rise in intracellular Ca^{2+} is a hallmark of environmental stress in the mammalian cardiomyocyte, and this leads to Ca^{2+} overload and the stimulation of cell death pathways. Interestingly, SR Ca^{2+} release is the major mechanism leading to Ca^{2+} overload, which suggests that ectothermic cardiomyocytes may be inherently protected from Ca^{2+} overload by possessing an SR, whose Ca^{2+} stores are harder to release. If intracellular Ca^{2+} levels rise to unusual levels in ectothermic myocytes during environmental perturbation, the SR could act as a Ca^{2+} buffer with a vast storage capacity. These concepts—although attractive—have not been supported so far by experimental evidence. Early studies using pharmacological inhibition of the ectothermic SR during acidosis or anoxia suggested that SR does not contribute to cardiac function during these insults. However, pharmacological inhibition can lead to compensation from other transporting mechanisms, such as the NCX.; so these results do not exclude the possibility of a role for the SR. In this regard, future studies should elucidate the role of the SR during environmental stress in ectothermic cardiomyocytes, where transport pathways can be isolated more accurately. This would be especially relevant for reptiles, such as freshwater turtles, which endure several months of anoxia and severe acidosis.

The Role of the SR in Avian, Neonatal, and Mammalian Atrial Cardiomyocytes

Myocytes from avian and neonatal hearts, and the mammalian atria appear to represent a *continuum* of structural and functional SR intermediates, between mammalian ventricular myocytes and ectotherm myocytes.

In mammalian atrial and avian myocytes, the SR plays a central role in E-C coupling, and CICR occurs despite the reduced or complete lack of a t-tubular system. Atrial myocytes that do possess a t-tubular network are from larger mammals,that have larger myocytes. Neonatal mammalian myocytes are devoid of t-tubules and have limited SR function, but both develop during ontogeny, coincident with myocyte growth.

The organization of CRUs and the short diffusional distance for Ca^{2+} transport in narrow cells allows for strong and fast contractions in avian myocytes and mammalian atrial myocytes. In atrial cells from small mammals, the jSR couples predominantly with the surface sarcolemma, leading to extensive and propagative CICR between neighboring CRUs, at the cell periphery.

The peripheral signal can propagate centripetally to activate the more extensive nonjunctional cSR in these cells and act as a Ca^{2+} signal accelerator and amplifier in the absence of, or a reduction in, the t-tubular system.

However, the peripheral Ca^{2+} signal does not always spread to more central CRUs, because the Ca^{2+} wave can be attenuated by distance and/or cellular constituents, that buffer Ca^{2+}, such as the mitochondria and the Ca^{2+} pumps of the fSR. Activation of nonjunctional cSR RyRs deep inside the mammalian atrial myocyte, which in turn activate the more centrally located myofilaments, can be achieved in much the same way as in ectotherms *via* pharmacological activators or sympathetic stimulation. When enhanced atrial contraction is required for ventricular filing, such as during exercise or stress, greater SR Ca^{2+} cycling is recruited and CICR propagates throughout the myocyte.

Birds are endothermic, with hearts rates and cardiac output pressures as high, if not higher, than mammals. However, bird cardiomyocytes are also thin and lack t-tubules (Table 2). [H3]ryanodine binding studies in pigeon and finch heart show the density of RyRs and the Ca^{2+} sensitivity of RyRs are similar to mammals. EM studies of the avian myocardium (i.e., finch, hummingbird and chick) show a high density of peripheral couplings containing CRUs [445]. However, the distance between these couplings may be too great for effective lateral transmission of CICR, suggesting peripheral RyR clusters are independently activated by LTCCs [445] in a manner similar to ectotherms.

However, unlike ectotherms, and akin to mammalian atrial myocytes, bird hearts also contain large quantities of cSR [445]. Importantly, CRUs in the cSR of birds are in closer proximity, than those found in the peripheral couplings (Table 3), which may enable propagated CICR release into the cell interior, facilitating strong and rapid contractions [445]. Functional studies of temporal and spatial Ca^{2+} flux, which would support this idea, are still lacking for bird myocardium. However, birds with faster heart rates (finch) have a higher density of RyRs and possess CRUs that are in closer proximity to each other than birds with slower heart rates (chicken) (Table 3).

This reinforces the connection between structural organization of the myocyte, the SR CRUs, and the strength and rate of cardiac contraction across vertebrate classes [476-479].

In summary, all vertebrates regulate the strength and rate of cardiac contraction by altering cellular Ca^{2+} flux in their myocytes. However, the mechanisms by which Ca^{2+} flux is altered varies with species, developmental stage, and cardiac tissue. SR Ca^{2+} cycling predominates in endothermic animals, and this facilitates high heart rates and robust cardiac pressure development. Ectotherms have thinner myocytes with higher SA-to-V ratios, which increases the efficacy of sarcolemmal Ca^{2+} flux in the absence of cSR and a t-tubular system. This arrangement contributes to the lower maximal heart rates and pressure development of many ectotherm species.

CRUs at peripheral couplings arepresent in all myocytes, but vary in size and frequency of distribution. The ability of a myocyte to release Ca^{2+} stored in the SR is related to the structural organization of CRUs, rather than the quantity of Ca^{2+} in the SR stores. This can be appreciated by the large ectothermic SR Ca^{2+} stores that are not routinely utilized for e-c coupling. The comparisonof SR function across vertebratessuggests that the SR basal role was Ca^{2+} storage, rather than Ca^{2+} release.

The mechanisms underlying the enhanced storage capacity of ectotherms are still being elucidated, but they are likely to involve different SR luminal regulation, enhanced luminal buffering, and RyRs that are less prone to activation. Correlative evidence suggests the structural organization required for CICR evolved in response to the demand for faster heart rates and/or stronger contractile force. This is true even of basal vertebrates, such as the cyclostome lamprey, which is a voracious predator during the parasitic life cycle phase and has recently been shown to utilize Ca^{2+} stores of the SR routinely for e-c coupling.

Most ectothermic species do not utilize the SR for Ca^{2+} release during e-c coupling, although some are able to overcome structural limitations and recruit SR Ca^{2+} cycling to augment trans-sarcolemmal flux and elevate cardiac performance, when required. The advantage conferred by increased control and rapidity of contractile initiation has been posited in the evolution of the skeletal form of e-c coupling, and the divergence of skeletal and cardiac e-c coupling from a common mechanism during vertebrate evolution.

Considering that the majority of extant vertebrates are ectothermic, more research into correlations between structure/function in these animals is expected to provide valuable insight into the role of the SR, in the evolution of vertebrate cardiac design.

References

[1] Barnard RJ, Edgerton VR, Furukawa T, Peter JB. Histochemical, biochemical, and contractile properties of red, white, and intermediate fibers. *Am J Physiol.* 1971; 220:410–414.

[2] Schiaffino S, Reggiani C. Fiber types in mammalian skeletal muscles. *Physiol Rev.* 2011; 91:1447–1531.

[3] Geiger PC, Cody MJ, Macken RL, Bayrd ME, Fang YH, Sieck GC. Mechanisms underlying increased force generation by rat diaphragm muscle fibers during development. *J Appl Physiol.* 2001; 90:380–388.

[4] Sieck GC, Zhan WZ, Prakash YS, Daood MJ, Watchko JF. SDH and actomyosin ATPase activities of different fiber types in rat diaphragm muscle. *J Appl Physiol.* 1995; 79:1629–1639.

[5] Sieck GC, Regnier M. Invited Review: plasticity and energetic demands of contraction in skeletal and cardiac muscle. *J Appl Physiol.* 2001; 90:1158–1164.

[6] Watchko JF, Daood MJ, Sieck GC. Myosin heavy chain transitions during development. Functional implications for the respiratory musculature. *Comp Biochem Physiol B Biochem Mol Biol.* 1998; 119:459–470.

[7] Geiger PC, Cody MJ, Han YS, Hunter LW, Zhan WZ, Sieck GC. Effects of hypothyroidism on maximum specific force in rat diaphragm muscle fibers. *J Appl Physiol.* 2002; 92:1506–1514.

[8] Gosselin LE, Johnson BD, Sieck GC. Age-related changes in diaphragm muscle contractile properties and myosin heavy chain isoforms. *Am J Respir Crit Care Med.* 1994; 150:174–178.

[9] Ben Hamida C, Soussi-Yanicostas N, Butler-Browne GS, Bejaoui K, Hentati F, Ben Hamida M. Biochemical and immunocytochemical

analysis in chronic proximal spinal muscular atrophy. *Muscle Nerve.* 1994; 17:400–410.

[10] Baloh RH, Rakowicz W, Gardner R, Pestronk A. Frequent atrophic groups with mixed-type myofibers is distinctive to motor neuron syndromes. *Muscle Nerve*. 2007; 36:107–110.

[11] Sieck GC. Physiological effects of diaphragm muscle denervation and disuse. Clin Chest Med. 1994; 15:641–659.

[12] Mantilla CB, Sieck GC. Invited review: Mechanisms underlying motor unit plasticity in the respiratory system. *J Appl Physiol*. 2003; 94:1230–1241.

[13] Fournier M, Sieck GC. Mechanical properties of muscle units in the cat diaphragm. *J Neurophysiol.* 1988; 59:1055–1066.

[14] Enad JG, Fournier M, Sieck GC. Oxidative capacity and capillary density of diaphragm motor units. *J Appl Physiol*. 1989; 67:620–627.

[15] Sieck GC, Fournier M, Prakash YS, Blanco CE. Myosin phenotype and SDH enzyme variability among motor unit fibers. *J Appl Physiol*. 1996; 80:2179–2189.

[16] Zhan WZ, Miyata H, Prakash YS, Sieck GC. Metabolic and phenotypic adaptations of diaphragm muscle fibers with inactivation. *J Appl Physiol.* 1997; 82:1145–1153.

[17] Miyata H, Zhan WZ, Prakash YS, Sieck GC. Myoneural interactions affect diaphragm muscle adaptations to inactivity. *J Appl Physiol*. 1995; 79:1640–1649.

[18] Geiger PC, Bailey JP, Zhan WZ, Mantilla CB, Sieck GC. Denervation-induced changes in myosin heavy chain expression in the rat diaphragm muscle. *J Appl Physiol*. 2003; 95:611–619.

[19] Hellyer NJ, Mantilla CB, Park EW, Zhan WZ, Sieck GC. Neuregulin-dependent protein synthesis in C2C12 myotubes and rat diaphragm muscle. *Am J Physiol Cell Physiol*. 2006; 291:C1056–1061.

[20] Mantilla CB, Sieck GC. Trophic factor expression in phrenic motor neurons. *Respir Physiol Neurobiol*. 2008; 164:252–262.

[21] Loeb JA, Hmadcha A, Fischbach GD, Land SJ, Zakarian VL. Neuregulin expression at neuromuscular synapses is modulated by synaptic activity and neurotrophic factors. *J Neurosci.* 2002; 22:2206–2214.

[22] Littlefield RS, Fowler VM. Thin filament length regulation in striated muscle sarcomeres: pointedend dynamics go beyond a nebulin ruler. *Semin Cell Dev Biol*. 2008; 19:511–519.

[23] Millman BM. The filament lattice of striated muscle. *Physiol Rev*. 1998; 78:359–391.

[24] Ingalls CP, Warren GL, Armstrong RB. Dissociation of force production from MHC and actin contents in muscles injured by eccentric contractions. *J Muscle Res Cell Motil*. 1998; 19:215–224.

[25] Yates LD, Greaser ML. Quantitative determination of myosin and actin in rabbit skeletal muscle. *J Mol Biol*. 1983; 168:123–141.

[26] Ryder JW, Lau KS, Kamm KE, Stull JT. Enhanced skeletal muscle contraction with myosin light chain phosphorylation by a calmodulin-sensing kinase. *J Biol Chem*. 2007; 282:20447–20454.

[27] Ottenheijm CA, van Hees HW, Heunks LM, Granzier H. Titin-based mechanosensing and signaling: role in diaphragm atrophy during unloading? *Am J Physiol Lung Cell Mol Physiol*. 2010; 300:L161–166.

[28] Bonnemann CG, Laing NG. Myopathies resulting from mutations in sarcomeric proteins. *Curr Opin Neurol*. 2004; 17:529–537.

[29] Prado LG, Makarenko I, Andresen C, Kruger M, Opitz CA, Linke WA. Isoform diversity of giant proteins in relation to passive and active contractile properties of rabbit skeletal muscles. *J Gen Physiol*. 2005; 126:461–480.

[30] Ottenheijm CA, Granzier H. New insights into the structural roles of nebulin in skeletal muscle. *J Biomed Biotechnol*. 2010; 2010:968139.

[31] Rowley KL, Mantilla CB, Ermilov LG, Sieck GC. Synaptic vesicle distribution and release at rat diaphragm neuromuscular junctions. *J Neurophysiol*. 2007; 98:478–487.

[32] Mantilla CB, Rowley KL, Fahim MA, Zhan WZ, Sieck GC. Synaptic vesicle cycling at typeidentified diaphragm neuromuscular junctions. *Muscle Nerve*. 2004; 30:774–783.

[33] Day NC, Wood SJ, Ince PG, Volsen SG, Smith W, Slater CR, Shaw PJ. Differential localization of voltage-dependent calcium channel alpha1 subunits at the human and rat neuromuscular junction. *J Neurosci*. 1997; 17:6226–6235.

[34] Ermilov LG, Mantilla CB, Rowley KL, Sieck GC. Safety factor for neuromuscular transmission at type-identified diaphragm fibers. *Muscle Nerve*. 2007; 35:800–803.

[35] Johnson BD, Sieck GC. Differential susceptibility of diaphragm muscle fibers to neuromuscular transmission failure. *J Appl Physiol*. 1993; 75:341–348.

[36] Bennett, H. S., and K. R. Porter. 1953. An electron microscopic study of sectioned breast muscle of the domestic fowl. *Amer. J. Anat*. 93:61.

[37] Porter, K. R., and G. E. Palade. 1957. Studies on the endoplasmic reticulum. III. Its form and distribution in striated muscle cells. J. 102. *Biophys. Biochem. Cytol.* 3:269.

[38] Porter, K. R. 1961. The sarcoplasmic reticulum: its recent history and present status. J. 101. *Biophys. Biochem. Cytol.* 10 (4, Suppl):211.

[39] Revel, J. P. 1962. The sarcoplasmic reticulum of the bat cricothyroid muscle. *J. Cell Biol.* 12:571.

[40] Endo, M. 1964. Entry of a dye into the sarcotubular system of muscle. *Nature.* 202:1115.

[41] Franzini-Armstrong, C. 1963. Sarcolemmal invaginations and the T-system in skeletal muscle fibers. *J. Cell Biol.* 19:24A. (Abstr.)

[42] Franzini-Armstrong, C., and K. R. Porter. 1964. Sarcolemmal invaginations and the T system in fish skeletal muscle. *Nature.* 202:355.

[43] Huxley, H. E. 1964. Evidence for continuity between the central elements of the triads and extracellular space in frog sartorius muscle. *Nature.* 202:1067.

[44] Page, E., and A. K. Solomon. 1960. Cat heart muscle in vitro. I. Cell volumes and intracellular concentrations in papillary muscle. *J. Gen. Physiol.* 44:327.

[45] Page, S. 1964. The organization of the sarcoplasmic reticulum in frog muscle. *J. Physiol.* (London). 175:10P.

[46] Page, S. G. 1965. A comparison of the fine structures of frog slow and twitch muscle fibres. *J. Cell Biol.* 26:477.

[47] Peachey, L. D. 1965. The sarcoplasmic reticulum and transverse tubules of the frog's sartorius. *J. Cell Biol.* 25:209.

[48] Gage, P. W., and R. S. Eisenberg. 1967. Action potentials without contraction in frog skeletal muscle fibers with disrupted transverse tubules. *Science.* 158:1702.

[49] Huxley, A. F., and R. E. Taylor. 1958. Local activation of striated muscle fibers. *J. Physiol.* (London). 144:426.

[50] Peachey, L., and K. R. Porter. 1959. Intracellular impulse conduction in muscle cells. *Science.* 129:721.

[51] Costantin, L. L., C. Franzini-Armstrong, and R. J. Podolsky. Localization of calcium accumulating structures in striated muscle. *Science.* 147:158.

[52] Ebashi, S., and F. Lipmann. 1962. Adenosine triphosphate-linked concentration of calcium ions in a particulate fraction of rabbit muscle. *J. Cell Biol.* 14:389.

[53] Hasselbach, W. 1964. Relaxation and the sarcotubular calcium pump. *Fed. Proc.* 23:909.

[54] Porter, K. R., and G. E. Palade. 1957. Studies on the endoplasmic reticulum. III. Its form and distribution in striated muscle cells. *J. Biophys. Biochem.* Cytol. 3:269.

[55] Fawcett, D. W. 1961. The sarcoplasmic reticulum of skeletal and cardiac muscle. *Circulation.* 24 (pt 2):336.

[56] Lindner, E. 1957. Die submikroskopische Morphologie des Herzmuskels. Z. Zellforsch. 45: 702.

[57] Battig, C. G., and F. N. Low. 1961. The ultrastructure of human cardiac muscle and its associated tissue space. *Amer. J. Anat.* 108:199.

[58] Fawcett, D. W. 1965. Observations on the T system and the cell-to-cell contacts of cardiac muscle. *Proc. 8th Intern. Anat. Congr.* Wiesbaden.

[59] Forssmann, W. G. 1965. Die dreidimensionale Ultrastructur des retikulosarkoplasmatischen Systems in Herzmuskel der Ratte and sein BeziehungzurphysiologischenFunktion. Proc. 8th Intern. *Anot. Congr.* pp. 37-38.

[60] Nelson, D. A., and E. S. Benson. 1963. On the structural continuities of the transverse tubular system of rabbit and human myocardial cells. *J. Cell Biol.* 16:297.

[61] Simpson, F. O., and S. J. Oertelis. 1962. The fine structure of sheep myocardial cells ; sarcolemmal invaginations and the transverse tubular system. *J. Cell Biol.* 12:91.

[62] Rayns, D. G., F. E. Simpson, and W. S. Bertrand. 1967. Transverse tubule apertures in mammalian myocardial cells: surface array. *Science.* 156:656.

[63] Simpson, F. O., and S. J. Oertelis. 1961. Relationship of the sarcoplasmic reticulum to the sarcolemma in sheep cardiac muscle. *Nature.* 189:758.

[64] Veratti, E. 1961. Investigations on the fine structure of striated muscle fibers (transla-tion). *J. Biophys. Biochem. Cytol.* 10 (No. 4, Suppl.): 1.

[65] Revel, J. P. 1962. The sarcoplasmic reticulum of the bat cricothyroid muscle. *J. Cell Biol.* 12:571.

[66] Nelson, D. A., and E. S. Benson. 1963. On the structural continuities of the transverse tubular system of rabbit and human myocardial cells. *J. Cell Biol.* 16:297.

[67] Simpson, F. O., and D. G. Rayns. 1968. The relationship between the transverse tubular system and other tubules at the Z disc levels in myocardial cells of the ferret. *Amer. J. Anat.* 122:193.

[68] Muir, A. R. 1967. The effects of divalent cations on the ultrastructure of the perfused rat heart. *J. Anat.* (London). 101:239.

[69] Nelson, D. A., and E. S. Benson. 1963. On the structural continuities of the transverse tubular system of rabbit and human myocardial cells. *J. Cell Biol.* 16:297.

[70] Dallner, G., P. Siekevitz, and G. E. Palade. 1966. Biogenesis of endoplasmic reticulum membranes. I. Structural and chemical differentiation in the developing rat hepatocyte. *J. Cell Biol.* 30:73.

[71] McNuTT, N. S., and D. W. Fawcett. 1969. The ultrastructure of the cat myocardium. II. *Atrial muscle. J. Cell Biol.* 41:46.

[72] Franzini-Armstrong, C. & A. O. Jorgenson. 1994. Structure and development of e-e coupling units in skeletal muscle. *Annu. Rev. Physiol.* 56: 509-534.

[73] Fwcher, B. E. & C. Franzini-Armstrong. 1996. Formation of junctions involved in excitation-contraction coupling in skeletal and cardiac muscle. *Proc. Nati. Acad. Sci. USA* 93:265-278.

[74] Stern, M. D., P. Gonzalo & E. R10s. 1997. A locai contro! model of excitation-contraction coupling in skeletal muscle. *J. Gen. Physiol.* 110: 1-26.

[75] Franzini-Armstrong, C. & C. W. Klstt. 1995. Alternate disposition of tetrads in periph- eral couplings of skeletal muscle. *J. Muscle Res. Celi Moti!.* 16: 19-324.

[76] Protasi, F., C. Franzini-Armstrong & B. Fwcher. 1997. Coordinated incorporation of skeletal muscle dihydropyridine receptors and ryanodine receptors in peripheral couplings of BC3Hl cells. *J. Celi* Biol. 137: 859-870.

[77] R10s, E. & G. Brum. 1987. Involvement of dihydropyridine receptors in excitation- con- traction coupling in skeletal muscle. *Nature* 325: 717-720.

[78] Tanabe, T. et al. 1988. Restoration of excitation-contraction coupling and slow calcium current in dysgenic muscle by dihydropyridine receptor complementary DNA. *Nature* 346: 567-569.

[79] Schneider, M. F. & W. K. Chandler. 1973. Voltage dependent charge movement of skele- tal muscle: A possible step in excitation-contraction coupling. *Nature* 242: 244-246.

[80] Jong, D. S., P. C. Pape & K. W. Chandler. 1995. Effect of sarcoplasmic reticulum calcium depletion on intramembranous charge movement in frog cut muscle fibers. *J. Gen. Phys- iol.* 106: 659-704.

[81] Tanabe, T. et al. 1990. Cardiac-type excitation-contraction coupling in dysgenic skeletal muscle injected with cardiac dihydropyridine receptor cDNA. *Nature* 344: 451-453.
[82] Nakai, J. et al. 1997. Functional non-equality of the cardiac and skeletal ryanodine recep- tors. *Proc. Nati. Acad. Sci. USA* 94: 1019-1022.
[83] Nakai, T. J. et al. 1996. Enhanced dihydropyridine receptor channel activity in the presence of ryanodine receptor. *Nature* 380: 72-75.
[84] Carl, S. L. et al. 1995. Immunolocalization of sarcolemmal dihydropyridine receptor and sarcoplasmic reticular triadin and ryanodine receptor in rabbit ventricle and atrium. *J. Celi Biol.* 129: 672-682.
[85] Protasi, F., X. -H. Sun & C. Franzini-Armstrong. 1996. Formation and maturation of calcium release units in developing and adult avian myocardium. *Dev. Biol.* 173:265-278.
[86] Sun, X. -H. et al. 1995. Molecular architecture of membranes involved in excitation- contraction coupling of cardiac muscle. *J. Celi Biol.* 129: 659-673.
[87] Eastwood, A. B., C. Franzini-Armstrong & C. Peracchia. 1982. Structure of mem- branes in crayfish muscle: Comparison of phasic and tonic fibers. *J. Muscle Res. Celi Motil.* 3:273-294.
[88] Palade, P. & S. Gvorke. 1993. Excitation-contraction coupling in crustacea: Do studies on these primitive creatures offer insights about EC coupling more generally? *J Muscle Res. Celi Motil.* 14: 283-287.
[89] Gvorke, S. & P. Palade. 1993. Role of locai Ca^{2+} domains in activation of Ca2+ induced Ca2+ release in crayfish muscle fibers. *Am. J. Phys.* 264: CI505-1512.
[90] Jewett, P. H., S. D. Leonaro & J. R. Sommer. 1973. Chicken cardiac muscle: Its elusive ex- tended junctional sarcoplasmic reticulum and sarcoplasmic reticulum fenestrations. *J. Celi Biol.* 56: 595-600.
[91] Sommer, J. R. 1995. Comparative anatomy: In praise of a powerful approach to elucidate mechanisms translating cardiac excitation into purposeful contraction. *J. Mol. Celi. Car- diol.* 27: 19-35.
[92] Junker, J., J. R. Sommer & G. Meissner. 1994. Extended junctional sarcoplasmic reticu- lum of avian cardiac muscle contains functional ryanodine receptors. *J. Biol. Chem.* 269: 1627-1634.
[93] Stern, M. D. 1992. Theory of excitation-contraction coupling in cardiac muscle. *Biophys. s. J.* 63:497-517.
[94] Franzini-Aamstrong, C., C. Champ & D. G. Ferguson. 1988. Discrimination between fast and slow twitch fibres of guinea pig skeletal

muscle using the relative surface density of junctional transverse tubule membrane. *J. Muscle Res. Celi Motil.* 9: 403-414.

[95] Appelt, D. et al. 1989. Quantitation of feet content in two types of muscle fibers from hind limb of the rat. *Tissue & Celi* 21:783-794.

[96] Appelt, D., V. Shen & C. Franzini-Armstrong. 1991. Quantitation of Ca ATPase, feet and mitochondria in super fast muscle fibres from the toadfish, Opsanus tau. *J. Muscle Res. Celi Motil.* 12: 543-552.

[97] Rome, L. C. et al. 1996. The whistle and the rattle: The design of sound producing muscles. *Proc. Nati. Acad. Sci. USA* 93:8095-8100.

[98] Devine CE, Somlyo AV, Somlyo AP. Sarcoplasmic reticulum and excitation-contraction coupling in mammalian smooth muscles. *J Cell Biol* 52: 690–718, 1972.

[99] Kuo KH, Herrera AM, Seow CY. Ultrastructure of airway smooth muscle. *Respir Physiol Neurobiol* 137: 197–208, 2003.

[100] Gabella G. Structure of smooth muscles. In: Pharmacology of Smooth Muscle, edited by Szekeres L and Papp LG. Berlin: Springer- Verlag, 1994.

[101] Sweeney M, Jones CJ, Greenwood SL, Baker PN, Taggart MJ. Ultrastructural features of smooth muscle and endothelial cells of isolated isobaric human placental and maternal arteries. *Placenta* 27: 635–647, 2006.

[102] Tasker PN, Michelangeli F, Nixon GF. Expression and distribution of the type 1 and type 3 inositol 1,4,5-trisphosphate receptor in developing vascular smooth muscle. *Circ Res* 84: 536–542, 1999.

[103] Bond M, Kitazawa T, Somlyo AP, Somlyo AV. Release and recycling of calcium by the sarcoplasmic reticulum in guinea-pig portal vein smooth muscle. *J Physiol* 355: 677–695, 1984.

[104] Nixon GF, Migneri GA, Somlyo AV. Immunogold localisation of inositol 1,4,5-triphosphate receptors and characterisation of ultrastructural features of the sarcoplasmic reticulum in phasic and tonic smooth muscle. *J Muscle Res Cell Motil* 15: 682–699, 1994.

[105] Gabella G. Structure of smooth muscle. In: Smooth Muscle: An Assessment of Current Knowledge, edited by Bulbring E, Brading AF, Jones AW and Tomita T. London: Arnold, 1981.

[106] Gabella G. Caveolae intracellulaires and sarcoplasmic reiculum in smooth muscle. *J Cell Sci* 8: 601–609, 1971.

[107] Peachey LD, Porter KR. Intracellular impulse conduction in smooth muscle cells. *Science* 129: 721–722, 1959.

[108] Prosser CL, Burnstock G, Kahn J. Conduction in smooth muscle: comparative structural properties. *Am J Physiol* 199: 545–552, 1960.

[109] Ross R, Klebanoff SJ. Fine structural changes in uterine smooth muscle and fibroblasts in response to estrogen. *J Cell Biol* 32: 155–167, 1967

[110] Somlyo AP. Excitation-contraction coupling and the ultrastructure of smooth muscle. *Circ Res* 57: 497–507, 1985.

[111] Huddart H, Hunt S. Visceral Muscle: Its Structure and Function. Glasgow, Scotland: Blackie, 1975.

[112] Paul AC (2001) Muscle length affects the architecture and pattern of innervation differently in leg muscles of mouse, guinea pig, and rabbit compared to those of human and monkey muscles. *Anat Rec* 262:301–309.

[113] Grounds MD, Shavlakadze T (2011) Growing muscle has different sarcolemmal properties from adult muscle: a proposal with scientific and clinical implications. *BioEssays* 33:458–468.

[114] Towler MC, Kaufman SJ, Brodsky FM (2004) Membrane traffic in skeletal muscle. *Traffic* 5:129–139.

[115] Mauro A (1961) Satellite cell of skeletal muscle fibers. *J Biophys Biochem Cytol* 9:493–495

[116] Snow MH (1977) The effects of aging on satellite cells in skeletal muscles of mice and rats. *Cell Tissue Res* 185:399–408.

[117] Ishikawa H (1966) Electron microscopic observations of satellite cells with special reference to the development of mammalian skeletal muscles. *Z Anat Entwicklungsgesch* 125:43–63.

[118] Muir AR, Kanji AH, Allbrook D (1965) The structure of the satellite cells in skeletal muscle. *J Anat* 99:435–444.

[119] Buckingham M, Relaix F (2007). The role of Pax genes in the development of tissues and organs: pax3 and Pax7 regulate muscle progenitor cell functions. *Annu Rev Cell Dev Biol* 23:645–673.

[120] Krauss RS (2010) Regulation of promyogenic signal transduction by cell–cell contact and adhesion. *Exp Cell Res* 316:3042–3049.

[121] Burattini S, Ferri P, Battistelli M et al. (2004) C2C12 murine myoblasts as a model of skeletal muscle development: morphofunctional characterization. *Eur J Histochem* 48:223–233.

[122] Tajika Y, Takahashi M, Khairani AF et al. (2014) Vesicular transport system in myotubes: ultrastructural study and signposting with vesicle-associated membrane proteins. *Histochem Cell Biol* 141:441–454.

[123] Hong W (2005) SNAREs and traffic. *Biochim Biophys Acta* 1744:120–144. doi:10. 1016/j. bbamcr. 2005. 03. 014.

[124] Grosshans BL, Ortiz D, Novick P (2006) Rabs and their effectors: achieving specificity in membrane traffic. *Proc Natl Acad Sci USA* 103:11821–11827. doi:10. 1073/pnas. 0601617103
[125] Nori A, Bortoloso E, Frasson F et al. (2004) Vesicle budding from endoplasmic reticulum is involved in calsequestrin routing to sarcoplasmic reticulum of skeletal muscles. *Biochem J* 379:505–512.
[126] Bach A-S, Enjalbert S, Comunale F et al. (2010) ADP-ribosylation factor 6 regulates mammalian myoblast fusion through phospholipase D1 and phosphatidylinositol 4,5-bisphosphate signaling pathways. *Mol Biol Cell* 21:2412–2424.
[127] Chen EH, Pryce BA, Tzeng JA et al. (2003) Control of myoblast fusion by a guanine nucleotide exchange factor, loner, and its effector ARF6. *Cell* 114:751–762.
[128] Klip A, Sun Y, Chiu TT, Foley KP (2014) Signal transduction meets vesicle traffic: the software and hardware of GLUT4 translocation. *AJP Cell Physiol* 306:C879–C886.
[129] Sanes JR, Lichtman JW (2001) Induction, assembly, maturation and maintenance of a postsynaptic apparatus. *Nat Rev Neurosci* 2:791–805.
[130] Burkin DJ, Kaufman SJ (1999) The alpha7beta1 integrin in muscle development and disease. *Cell Tissue Res* 296:183–190.
[131] Neville C, Rosenthal N, McGrew M et al. (1997) Skeletal muscle cultures. *Methods Cell Biol* 52:85–116.
[132] Schneider, M. F. (1981) *Annu. Rev. Physiol.* 43, 509-517.
[133] Schneider, M. F. & Chandler, W. K. (1973) *Nature* (London) 242, 244-246.
[134] Rios, E. & Brum, G. (1987) *Nature* (London) 325, 717-720.
[135] Beam, K. G., Knudson, C. M. & Powell, J. A. (1986) Nature (London) 320, 168-170.
[136] Fleischer, S. & Inui M. (1989) *Annu. Rev. Biophys. Chem.* 18,333-365.
[137] Meissner, G. (1994) *Annu. Rev. Physiol.* 56, 485-508.
[138] Melzer, W., Herrmann-Frank, A. & Luttgau, H. C. (1994) Biochim. *Biophys. Acta* 1241, 59-116.
[139] Nabauer, M., Gallewaert, G., Cleeman, L. & Morad, M. (1989) Science 244, 800-803.
[140] Fabiato, A. (1985) *J. Gen. Physiol.* 245, 291-320.
[141] Stern, M. D. & Lakatta, E. G. (1992) *FASEB J.* 6, 3092-3100.
[142] Cannell, M. B., Cheng, H. & Lederer, W. J. (1995) *Science* 268, 1045-1049.
[143] Franzini-Armstrong, C. (1970) *J. Cell Biol.* 47, 488-499.

[144] Ferguson, D. G., Schwartz, H. & Franzini-Armstrong, C. (1984) *J. Cell Biol.* 99, 1735-1742.

[145] Wagenknecht, T., Grassucci, R. Frank, J. Saito, A. Inui, M. & Fleischer, S. (1989) *Nature* (London) 338, 167-170.

[146] Radermacher, M., Rao, V., Grassucci, R., Frank, J., Timerman, A. P., Fleischer, S. & Wagenknecht, T. (1994) *J. Cell Biol.* 127, 411-423.

[147] Serysheva, I. I., Orlova, E. V., Chiu, W., Sherman, M. B., Hamilton, S. L. & van Heel, M. (1995) *Nat. Struct. Biol.* 2, 18-24.

[148] Bossen, E., Sommer, J. R. & Waugh, R. A. (1978) *Tissue Cell* 10, 773-784.

[149] Block, B. A., Leung, A., Campbell, K. P. & Franzini-Armstrong, C. (1988) *J. Cell Biol.* 107, 2587-2600.

[150] Jorgensen, A. O., Shen, A. C. -Y., Arnold, W., Leung, A. T. & Campbell, K. P. (1989) *J. Cell Biol.* 109, 135-147.

[151] Flucher, B. E., Morton, M. E., Froehner, S. C. & Daniels, M. P. (1990) *Neuron* 5, 339-351.

[152] Franzini-Armstrong, C., Pincon-Raymond, M. & Rieger, F. (1991) *Dev. Biol.* 146, 364-376.

[153] Takekura, H., Bennett, L., Tanabe, T., Beam, K, G. & Franzini-Armstrong, C. (1994) *Biophys. J.* 67, 793-804.

[154] Sun, X-H., Protasi, F., Takahashi, M., Takeshima, H., Ferguson, D. G. & Franzini-Armstrong, C. (1995) *J. Cell Biol.* 129,659-673.

[155] Pizarro, G., Csernoch, L., Uribe, I., Rodriguez, M. & Rios, E. (1991) *J. Gen. Physiol. 97*, 913-947.

[156] Pape, P. C., Jong, D. -S. & Chandler, W. K. (1995). *J. Gen. Physiol.* 106, 259-336.

[157] Caswell, A. H., Lau, Y. H., Garcia, M. & Brunschwig, J. -P. (1979) *J. Biol. Chem.* 254, 202-208.

[158] Takekura, H., Takeshima, H., Nishimura, S., Imoto, K., Takahashi, M., Tanabe, T., Flockerzi, V., Hofman, F. & Franzini-Armstrong, C. (1995) *J. Muscle Res. Cell Motil.* 16, 465-480.

[159] Tanabe, T., Beam, K. G. Adams, B. A., Nicodome, T. & Numa, S. (1990) *Nature* (London) 346, 567-569.

[160] Tanabe, T., Mikami, A., Numa, S. & Beam, K. G. (1990) *Nature* (London) 344, 451-453.

[161] Lu, X., Xu, L. & Meissner, G. (1994) *J. Biol. Chem.* 269, 6511-6516.

[162] el-Hayek, R., Antoniu, B., Wang, J., Hamilton, S. L. & Ikemoto, N. (1995) *J. Biol. Chem.* 270, 22116-22118.

[163] Nakai, J., Dirksen, R. T., Nguyen, H. T., Pessah, I. N., Beam, K. G. & Allen, P. D. (1996) *Nature* (London) 380, 72-74.

[164] Brandt, N. R., Caswell, A. H. Wen, S. R. & Talvenheimo, J. A. (1990) *J. Membr. Biol.* 113, 237-251.

[165] Kim, K. C., Caswell, A. H., Brunschwig, J. -P. & Brandt, N. R. (1990b) *J. Membr. Biol.* 113, 221-235.

[166] Caswell A. H., Brandt, N. R., Brunschwig, J. P. & Purkerson, S. (1991) *Biochemistry* 30, 7507-7513.

[167] Flucher, B. E., Andrews, S. B., Fleischer, S., Marks, A. R., Caswell, A. & Powell, J. A. (1993a) *J. Cell Biol.* 123, 1161-1174.

[168] Knudson, C. M., Stang, K. K., Moomaw, C. R., Slaughter, C. A. & Campbell, K. P. (1993) *J. Biol. Chem.* 268, 12646-12654.

[169] Lewis-Carl, S., Felix, K., Caswell, A. H. Brandt, N. R., Ball, W. J., Vaghy, P. L., Meissner, G. & Ferguson, D. G. (1995) *J. Cell Biol.* 129, 673-682.

[170] Yuan, S., Arnold, W. & Jorgensen, A. 0. (1991) *J. Cell Biol.* 112, 289-301.

[171] Flucher, B. E., Andrews, S. B. & M. P. Daniels. (1994) *Mol. Biol. Cell* 5, 1105-1118.

[172] Protasi, F., Sun, X. -H. & Franzini-Armstrong, C. (1996) *Dev. Biol.* 173, 265-278.

[173] Flucher, B. E. (1992) *Dev. Biol.* 154, 245-260.

[174] Flucher, B. E., Terasaki, M., Chin, H., Beeler, T. & Daniels, M. P. (1991a) *Dev. Biol.* 145, 77-90.

[175] Flucher, B. E., Phillips, J. L. & Powell, J. A. (1991) *J. Cell Biol.* 115, 1345-1356.

[176] Romey, G., Garcia, L., Dimitriadu, V., Pincon-Raymond, M., Rieger, F. & Lazdunski, M. (1989) *Proc. Natl. Acad. Sci. USA* 86, 2933-2937.

[177] Takekura, H., Sun, X. -H. & Franzini-Armstrong, C. (1994) *J. Muscle Res. Cell Motil.* 15, 102-118.

[178] Flucher, B. E., Takekura, H. & Franzini-Armstrong, C. (1993) *Dev. Biol.* 160, 135-147.

[179] Adams, B. & Beam, K. G. (1991) *FASEB J.* 4, 2809-2816.

[180] Rieger, F., Pincon-Raymond, M., Tassin, A. M., Garcia, L., Romey, G., Fosset, M. & Lazdunski, M. (1987) *Biochimie* 69, 411-417.

[181] Tanabe, T., Beam, K. G., Powell, J. A. & Numa, S. (1988) *Nature* (London) 336, 134-139.

[182] Adams, B. A., Tanabe, T., Mikami, A., Numa, S. & Beam, K. G. (1990) *Nature* (London) 346, 569-572.

[183] Chaudhari, N. (1992) *J. Biol. Chem.* 25, 25636-25639.

[184] Knudson, C. M., Chaudhari, N., Sharp, A. H., Powell, J. A., Beam, K. G. & Campbell, K. P. (1989) *J. Biol. Chem.* 264,1345-1348.

[185] Flucher, B. E., Phillips, J. L., Powell, J. A., Andrews, S. B. & Daniels, M. P. (1992) *Dev. Biol.* 155, 266-280.

[186] Pincon-Raymond, M., Rieger, F., Fosset, M. & Lazdunski, M. (1998) *Dev. Biol.* 112, 458-466.

[187] Courbin, P., Koenig, J., Ressouches, A., Beam, K. G. & Powell, J. A. (1989) *Neuron* 2, 1341-1350.

[188] Chaudhari, N. & Beam, K. G. (1993) *Dev. Biol.* 155, 507-515.

[189] Varadi, G., Mikala, G., Lory, P., Varadi, M., Drouet, B., Pingon-Raymond, M. & Schwartz, A. (1995) *FEBS Lett.* 368, 405-410.

[190] Adams, B. A. & Beam, K. G. (1989) *J. Gen Physiol.* 94,429-444.

[191] Powell, J. A., Petherbridge, L. & Flucher, B. E. (1996) *J. Cell Biol.*, in press.

[192] Dolber, P. C. & Sommer, J. R. (1984) *J. Ultrastructure Res.* 87, 190-196.

[193] Jewett, P. H., Sommer, J. R. & Johnson, E. A. (1971) *J. Cell Biol.* 59, 50-65.

[194] Jorgensen, A. O., Shen, A. C. -Y., Arnold, W., PcPhearson, P. S. & Campbell, K. P. (1993) *J. Cell Biol.* 120, 969-980.

[195] Junker, J., Sommer, J. R., Sar, M. & Meissner, G. (1994) *J. Biol. Chem.* 269, 1627-1634.

[196] Takeshima, H., Iino, M., Takekura, H., Nishi, M., Kuno, J., Minowa, O., Takano, H. & Noda, T. (1994) *Nature* (London) 369, 556-559.

[197] Takeshima, H., Yamazawa, T., Ikemoto. T., Takekura, H., Nishi, M., Noda, T. & lino, M. (1995) *EMBO J.* 14, 2999-3006.

[198] Takekura, H., Nishi, M., Toda, T., Takeshima, H. & Franzini-Armstrong, C. (1995) *Proc. Natl. Acad. Sci. USA* 92, 3381-3385.

[199] Edge, M. B. (1970) *Dev. Biol.* 23, 634-650.

[200] Marks, A. R., Taubman, M. B., Saito, A., Dai, Y. & Fleischer, S. (1991) *J. Cell Biol.* 114, 303-312.

[201] Flucher B, Phillips J, Powell J, Andrews S, Daniels M. Coordinated development of myofibrils, sarcoplasmic reticulum and transverse tubules in normal and dysgenic mouse skeletal muscle, in vivo and in vitro. *Dev Bio*. 1992;150:266-280.

[202] Adams BA, Beam KG. Muscular dysgenesis in mice: a model system for studying excitation-contraction coupling. *FASEB J.* 1990;4:2809-2816.

[203] Beam KG, Knudson M, Powell JA. A lethal mutation in mice eliminates the slow calcium current in skeletal muscle cells. *Nature.* 1986;320:168-170.

[204] Chang C, Gutierrez L, Mundina-Weilenmann C, Hosey M. Dihydropyridine-sensitive calcium channels from skeletal muscle. *JBiol Chem* 1991;266:16395-16400.

[205] Gutierrez LM, Brawley RM, Hosey MM. Dihydropyridinesensitive calcium channels from skeletal muscle, I: roles of subunits in channel activity. *J Biol Chem.* 1991;206:16387-16397.

[206] Tanabe T, Takeshima H, Mikami A, Flockerzi V, Takahashi H, Kangawa K, Kojima M, Matsuo H, Hirose T, Numa S. Primary structure of the receptor for calcium channel blockers from skeletal muscle. *Nature*. 1987;328:313-328.

[207] 8. Ellis SB, Williams ME, Ways NR, Brenner R, Sharp A, Leung AT, Campbell KP, McKenna E, Koch WJ, Hui A, Schwartz A, Harpold A. Sequence and expression of mRNAs encoding the α_1, and α_2 subunits of a DHP-sensitive calcium channel. *Nature.* 1988;241: 1661-1664.

[208] Jay SD, Ellis SB, McCue AF, Williams ME, Vedvick TS, Harpold MM, Campbell KP. Primary structure of the gamma subunit of the DHP-sensitive calcium channel from skeletal muscle. *Science*. 1990;248:490-492.

[209] Ruth P, Rohrkasten A, Biel M, Bosse E, Regulla S, Meyer H, Flockerzi V, Hofman F. Primary structure of the 13 subunit of the DHP-sensitive calcium channel from skeletal muscle. *Science.* 1989;245:1115-1118.

[210] Varadi G, Lory P, Schultz D, Varadi M, Schwartz A. Acceleration of activation and inactivation by the /-subunit of the skeletal muscle calcium channel. *Nature.* 1991;352:159-162.

[211] Lacerda A, Kim H, Ruth P, Perez-Reyes E, Flockerzi V, Hofmann F, Birnbaumer L, Brown A. Normalization of current kinetics by interaction between a, and /3 subunits of the skeletal muscle dihydropyridine-sensitive Ca'2 channel. *Nature.* 1991;352:527-530.

[212] Artman M, Graham T, Boucek R. Effects of postnatal maturation on myocardial contractile responses to calcium antagonists and changes in contraction frequency. *J Cardiovasc Pharmacol*. 1985;7: 850-855.

[213] Artman M. Developmental changes in myocardial contractile responses to inotropic agents. *Cardiovasc Res*. 1992;26:1-12.

[214] Chin T, Friedman W, Klitzner T. Developmental changes in cardiac myocyte calcium regulation. *Circ Res*. 1990;67:574-579.

[215] Huynh TV, Chen F, Wetzel GT, Friedman WF, Klitzner TS. Developmental changes in membrane Call and K' currents in fetal, neonatal, adult rabbit ventricular myocytes. *Circ Res.* 1992; 70:508-515.
[216] Klitzner, T Friedman W. Excitation-contraction coupling in developing mammalian myocardium: evidence from voltage clamp studies. *Pediatr Res.* 1988;23:428-432.
[217] Wetzel G, Chen F, Klitzer T. L- and T-type calcium channels in acutely isolated neonatal and adult cardiac myocytes. *Pediatr Res.* 1991;30:89-94.
[218] Jarmakani J, Nakanishi T, George B, Bers D. Effect of extracellular calcium on myocardial mechanical function in the neonatal rabbit. *Dev Pharmacol Ther.* 1982;5:1-13.
[219] Mahony L, Jones L. Developmental changes in cardiac sarcoplasmic reticulum in sheep. *J Biol Chem.* 1986;261:15257-15265.
[220] Seguchi M, Harding J, Jarmakani J. Developmental change in the function of sarcoplasmic reticulum. *J Mol Cell Cardiol.* 1986;18: 189-195.
[221] Tanaka H, Shigenobu K. Effect of ryanodine on neonatal and adult rat heart: developmental increase in sarcoplasmic reticulum function. *J Mol Cell Cardiol* 1989;21:1305-1313.
[222] 40. Seguchi M, Jarmakani J, George B, Harding J. Effect of Ca'2 antagonists on mechanical function in the neonatal heart. *Pediatr Res.* 1986;20:838-842.
[223] Arai M, Otsu K, MacLennan D, Periasamy M. Regulation of sarcoplasmic reticulum gene expression during cardiac and skeletal muscle development. *Am J Physiol.* 1992;262:C614-C620.
[224] Brillantes A, Allen P, Takahasi T, Izumo S, Marks A. Differences in cardiac calcium release channel (ryanodine receptor) expression in myocardium from patients with end-stage heart failure caused by ischemic versus dilated cardiomyopathy. *Circ Res.* 1992;71: 18-26.
[225] Giannini G, Clementi E, Ceci R, Marziali G, Sorrentino V. Expression of a ryanodine receptor-Ca 2+ channel that is regulated by TGF-,B. *Science.* 1992;257:91-93.
[226] Marks AR, Taubman M, Saito A, Dai Y, Fleischer S. The ryanodine receptor/junctional channel complex is regulated by growth factors in a myogenic cell line. *J Cell Biol.* 1991;114:303-312.
[227] Izumo S, Nadal-Ginard B, Mahdavi V. Protooncogene induction and reprogramming of cardiac gene expression produced by pressure overload. *Proc Natl Acad Sci USA.* 1988;85:339-343.

[228] Parker T, Packer S, Schneider M. Peptide growth factors can provoke "fetal" contractile protein gene expression in rat cardiac myocytes. *J Clin Invest.* 1990;85:507-514.
[229] Wetzel G, Chen F, Klitzer T. L- and T-type calcium channels in acutely isolated neonatal and adult cardiac myocytes. *Pediatr Res.* 1991;30:89-94.
[230] Fryer MW, Stephenson DG. Total and sarcoplasmic reticulum calcium contents of skinned fibres from rat skeletal muscle. *J Physiol* (Lond) 1996;493:357–370.
[231] Costello B, Chadwick C, Saito A, Chu A, Maurer A, Fleischer S. Characterization of the junctional face membrane from terminal cisternae of sarcoplasmic reticulum. *J Cell Biol* 1986;103:741–753.
[232] Beard NA, Laver DR, Dulhunty AF. Calsequestrin and the calcium release channel of skeletal and cardiac muscle. *Prog Biophys Mol Biol* 2004;85:33– 69.
[233] MacLennan DH, Wong PT. Isolation of a calcium-sequestering protein from sarcoplasmic reticulum. *Proc Natl Acad Sci USA* 1971;68:1231–1235.
[234] Arai M, Otsu K, MacLennan DH, Periasamy M. Regulation of sarcoplasmic reticulum gene expression during cardiac and skeletal muscle development. *Am J Physiol Cell Physiol* 1992; 262:C614–C620.
[235] Damiani E, Margreth A. Characterization study of the ryanodine receptor and of calsequestrin isoforms of mammalian skeletal muscles in relation to fibre types. *J Muscle Res Cell Motil* 1994;15:86 –101.
[236] Park KW, Goo JH, Chung HS, Kim H, Kim DH, Park WJ. Cloning of the genes encoding mouse cardiac and skeletal calsequestrins: expression pattern during embryogenesis. *Gene* 1998;217:25–30.
[237] Sacchetto R, Volpe P, Damiani E, Margreth A. Postnatal development of rabbit fast-twitch skeletal muscle: accumulation, isoform transition and fibre distribution of calsequestrin. *J Muscle Res Cell Motil* 1993;14:646–653.
[238] Ludolph DC, Konieczny SF. Transcription factor families: muscling in on the myogenic program. *FASEB J* 1995;9: 1595–1604.
[239] Wang S, Trumble WR, Liao H, Wesson CR, Dunker AK, Kang CH. Crystal structure of calsequestrin from rabbit skeletal muscle sarcoplasmic reticulum. *Nat Struct Biol* 1998;5: 476–483.
[240] Krause KH, Milos M, Luan-Rilliet Y, Lew DP, Cox JA. Thermodynamics of cation binding to rabbit skeletal muscle

calsequestrin. Evidence for distinct Ca2- and Mg2-binding sites. *J Biol Chem* 1991;266:9453–9459.

[241] Franzini-Armstrong C, Kenney LJ, Varriano-Marston E. The structure of calsequestrin in triads of vertebrate skeletal muscle: a deep-etch study. *J Cell Biol* 1987;105:49 –56.

[242] Fliegel L, Ohnishi M, Carpenter MR, Khanna VK, Reithmeier RA, MacLennan DH. Amino acid sequence of rabbit fast-twitch skeletal muscle calsequestrin deduced from cDNA and peptide sequencing. *Proc Natl Acad Sci USA* 1987;84:1167–1171.

[243] Zhang L, Kelley J, Schmeisser G, Kobayashi YM, Jones LR. Complex formation between junctin, triadin, calsequestrin, and the ryanodine receptor: proteins of the cardiac junctional sarcoplasmic reticulum membrane. *J Biol Chem* 1997; 272:23389–23397.

[244] Shin DW, Ma J, Kim DH. The asp-rich region at the carboxylterminus of calsequestrin binds to Ca^{2+} and interacts with triadin. *FEBS Lett* 2000;486:178 –182.

[245] Glover L, Culligan K, Cala S, Mulvey C, Ohlendieck K. Calsequestrin binds to monomeric and complexed forms of key calcium-handling proteins in native sarcoplasmic reticulum membranes from rabbit skeletal muscle. *Biochim Biophys Acta* 2001;1515:120–132.

[246] Herzog A, Szegedi C, Jona I, Herberg FW, Varsanyi M. Surface plasmon resonance studies prove the interaction of skeletal muscle sarcoplasmic reticular Ca2 release channel/ ryanodine receptor with calsequestrin. *FEBS Lett* 2000;472: 73–77.

[247] Brandt NR, Caswell AH, Wen SR, Talvenheimo JA. Molecular interactions of the junctional foot protein and dihydropyridine receptor in skeletal muscle triads. *J Membr Biol* 1990;113:237–251.

[248] Marty I, Thevenon D, Scotto C, Groh S, Sainnier S, Robert M, et al. Cloning and characterization of a new isoform of skeletal muscle triadin. *J Biol Chem* 2000;275:8206–8212.

[249] Thevenon D, Smida-Rezgui S, Chevessier F, Groh S, Henry-Berger J, Beatriz RN, et al. Human skeletal muscle triadin: gene organization and cloning of the major isoform, Trisk 51. *Biochem Biophys Res Commun* 2003;303:669–675.

[250] Vassilopoulos S, Thevenon D, Rezgui SS, Brocard J, Chapel A, Lacampagne A, et al. Triadins are not triad-specific proteins: two new skeletal muscle triadins possibly involved in the architecture of sarcoplasmic reticulum. *J Biol Chem* 2005;280:28601–28609.

[251] Knudson CM, Stang KK, Jorgensen AO, Campbell KP. Biochemical characterization of ultrastructural localization of a major junctional sarcoplasmic reticulum glycoprotein (triadin). *J Biol Chem* 1993;268:12637–12645.

[252] Kobayashi YM, Alseikhan BA, Jones LR. Localization and characterization of the calsequestrin-binding domain of triadin 1: evidence for a charged -strand in mediating the protein-protein interaction. *J Biol Chem* 2000;275:17639–17646.

[253] Lee JM, Rho SH, Shin DW, Cho C, Park WJ, Eom SH, et al. Negatively charged amino acids within the intraluminal loop of ryanodine receptor are involved in the interaction with triadin. *J Biol Chem* 2004;279:6994–7000.

[254] Jones LR, Zhang L, Sanborn K, Jorgensen AO, Kelley J. Purification, primary structure, and immunological characterization of the 26-kDa calsequestrin binding protein (junctin) from cardiac junctional sarcoplasmic reticulum. *J Biol Chem* 1995;270:30787–30796.

[255] Suk JY, Kim YS, Park WJ. HRC (histidine-rich Ca2 binding protein) resides in the lumen of sarcoplasmic reticulum as a multimer. Biochem *Biophys Res Commun* 1999;263:667–671.

[256] Hofmann SL, Goldstein JL, Orth K, Moomaw CR, Slaughter CA, Brown MS. Molecular cloning of a histidine-rich Ca2-binding protein of sarcoplasmic reticulum that contains highly conserved repeated elements. *J Biol Chem* 1989;264:18083–18090.

[257] Lee HG, Kang H, Kim DH, Park WJ. Interaction of HRC (histidine-rich Ca2-binding protein) and triadin in the lumen of sarcoplasmic reticulum. *J Biol Chem* 2001;276:39533–39538.

[258] Treves S, Feriotto G, Moccagatta L, Gambari R, Zorzato F. Molecular cloning, expression, functional characterization, chromosomal localization, and gene structure of junctate, a novel integral calcium binding protein of sarco(endo)plasmic reticulum membrane. *J Biol Chem* 2000;275:39555– 39568.

[259] Yoshida M, Minamisawa S, Shimura M, Komazaki S, Kume H, Zhang M, et al. Impaired Ca2 store functions in skeletal and cardiac muscle cells from sarcalumenin-deficient mice. *J Biol Chem* 2005;280:3500–3506.

[260] Lahat H, Pras E, Eldar M. A missense mutation in CASQ2 is associated with autosomal recessive catecholamine-induced polymorphic ventricular tachycardia in Bedouin families from Israel. *Ann Med* 2004;36(suppl 1):87–91.

[261] Fill M, Copello JA. Ryanodine receptor calcium release channels. *Physiol Rev* 2002;82:893–922.

[262] Dirksen RT, Avila G. Distinct effects on Ca2 handling caused by malignant hyperthermia and central core disease mutations in RyR1. *Biophys J* 2004;87:3193–3204.

[263] Sipido KR, Maes M, Van de WF. Low efficiency of Ca^{2+} entry through the Na^+-Ca^{2+} exchanger as trigger for Ca^{2+} release from the sarcoplasmic reticulum. A comparison between L-type Ca2þ current and reverse-mode Na^+-Ca^{2+} exchange. *Circ Res* 1997;81:1034–1044.

[264] Sipido KR, Carmeliet E, Van de WF. T-type Ca^{2+} current as a trigger for Ca^{2+} release from the sarcoplasmic reticulum in guinea-pig ventricular myocytes. *J Physiol* 1998;508:439–451.

[265] Scriven DR, Dan P, Moore ED. Distribution of proteins implicated in excitation-contraction coupling in rat ventricular myocytes. *Biophys J* 2000;79:2682–2691.

[266] Fabiato A. Time and calcium dependence of activation and inactivation of calcium-induced release of calcium from the sarcoplasmic reticulum of a skinned canine cardiac Purkinje cell. *J Gen Physiol* 1985;85:247–289.

[267] Satoh H, Delbridge LM, Blatter LA, Bers DM. Surface:volume relationship in cardiac myocytes studied with confocal microscopy and membrane capacitance measurements: species-dependence and developmental effects. *Biophys J* 1996;70:1494–1504.

[268] Brette F, Orchard CH. Density and sub-cellular distribution of cardiac and neuronal sodium channel isoforms in rat ventricular myocytes. *Biochem Biophys Res Commun* 2006;348:1163–1166.

[269] Brette F, Salle L, Orchard CH. Quantification of calcium entry at the T-tubules and surface membrane in rat ventricular myocytes. *Biophys J* 2006;90:381–389.

[270] Chen-Izu Y, McCulle SL, Ward CW, Soeller C, Allen BM, Rabang C, Cannell MB, Balke CW, Izu LT. Three-dimensional distribution of ryanodine receptor clusters in cardiac myocytes. *Biophys J* 2006;91:1–13.

[271] Inoue M, Bridge JH. Variability in couplon size in rabbit ventricular myocytes. *Biophys J* 2005;89:3102–3110.

[272] Franzini-Armstrong C, Protasi F, Ramesh V. Shape, size, and distribution of Ca^{2+} release units and couplons in skeletal and cardiac muscles. *Biophys J* 1999;77:1528–1539.

[273] Bers DM, Stiffel VM. Ratio of ryanodine to dihydropyridine receptors in cardiac and skeletal muscle and implications for E-C coupling. *Am J Physiol* 1993;264:C1587–C1593.
[274] Block BA, Imagawa T, Campbell KP, Franzini-Armstrong C. Structural evidence for direct interaction between the molecular components of the transverse tubule/sarcoplasmic reticulum junction in skeletal muscle. *J Cell Biol* 1988;107:2587–2600.
[275] Brette F, Salle L, Orchard CH. Differential modulation of L-type Ca^{2+} current by SR Ca^{2+} release at the T-tubules and surface membrane of rat ventricular myocytes. *Circ Res* 2004;95:e1–e7.
[276] Shacklock PS, Wier WG, Balke CW. Local Ca^{2+} transients (Ca^{2+} sparks) originate at transverse tubules in rat heartCells. *J Physiol* 1995;487: 601–608.
[277] Song LS, Wang SQ, Xiao RP, Spurgeon H, Lakatta EG, Cheng H. b-Adrenergic Stimulation Synchronizes Intracellular Ca2þ Release During Excitation-Contraction Coupling in Cardiac Myocytes. *Circ Res* 2001;88:794–801.
[278] Cordeiro JM, Spitzer KW, Giles WR, Ershler PE, Cannell MB, Bridge JH. Location of the initiation site of calcium transients and sparks in rabbit heart Purkinje cells. *J Physiol* 2001;531:301–314.
[279] Haddock PS, Coetzee WA, Cho E, Porter L, Katoh H, Bers DM, Jafri MS, Artman M. Subcellular $[Ca^{2+}]i$ gradients during excitation-contraction coupling in newborn rabbit ventricular myocytes. *Circ Res* 1999;85: 415–427.
[280] Huser J, Lipsius SL, Blatter LA. Calcium gradients during excitationcontraction coupling in cat atrial myocytes. *J Physiol* 1996;494:641–651.
[281] Louch WE, Bito V, Heinzel FR, Macianskiene R, Vanhaecke J, Flameng W, Mubagwa K, Sipido KR. Reduced synchrony of Ca^{2+} release with loss of T-tubules-a comparison to Ca^{2+} release in human failing cardiomyocytes. *Cardiovasc Res* 2004;62:63–73.
[282] Yang Z, Pascarel C, Steele DS, Komukai K, Brette F, Orchard CH. Na^+-Ca^{2+} exchange activity is localized in the T-tubules of rat ventricular myocytes. *Circ Res* 2002;91:315–322.
[283] Beuckelmann DJ, Wier WG. Mechanism of release of calcium from sarcoplasmic reticulum of guinea-pig cardiac cells. *J Physiol* 1988;405:233–255.

[284] Fabiato A. Simulated calcium current can both cause calcium loading in and trigger calcium release from the sarcoplasmic reticulum of a skinned canine cardiac Purkinje cell. *J Gen Physiol* 1985;85:291–320.

[285] Trafford AW, Diaz ME, Negretti N, Eisner DA. Enhanced Ca^{2+} current and dccrcascd Ca^{2+} cfflux rcstorc sarcoplasmic reticulum Ca^{2+} content after depletion. *Circ Res* 1997;81:477–484.

[286] Trafford AW, Diaz ME, Eisner DA. Stimulation of Ca-induced Ca^{2+} release only transiently increases the systolic Ca^{2+} transient: measurements of Ca^{2+} fluxes and sarcoplasmic reticulum Ca. *Cardiovasc Res* 1998;37: 710–717.

[287] Trafford AW, Diaz ME, Sibbring GC, Eisner DA. Modulation of CICR has no maintained effect on systolic Ca^{2+}: simultaneous measurements of sarcoplasmic reticulum and sarcolemmal Ca^{2+} fluxes in rat ventricular myocytes. *J Physiol* 2000;522:259–270.

[288] Eisner DA, Trafford AW, Diaz ME, Overend CL, O'Neill SC. The control of Ca^{2+} release from the cardiac sarcoplasmic reticulum: regulation versus autoregulation. *Cardiovasc Res* 1998;38:589–604.

[289] Orchard C, Brette F. T-tubules and sarcoplasmic reticulum function in cardiac ventricular myocytes. *Cardiovascular Research* (2008) 77, 237–244.

[290] Trafford AW, Diaz ME, O'Neill SC, Eisner DA. Comparison of subsarcolemmal and bulk calcium concentration during spontaneous calcium release in rat ventricular myocytes. *J Physiol* 1995;488:577–586.

[291] Despa S, Brette F, Orchard CH, Bers DM. Na/Ca exchange Na/K-ATPase function are equally concentrated in transverse tubules of rat ventricular myocytes. *Biophys J* 2003;85:3388–3396.

[292] Pasek M, Simurda J, Christe G. The functional role of cardiac T-tubules explored in a model of rat ventricular myocytes. *Philos Transact A Math Phys Eng Sci* 2006;364:1187–1206.

[293] Jurevicius J, Fischmeister R. cAMP compartmentation is responsible for a local activation of cardiac Ca^{2+} channels by beta-adrenergic agonists. *Proc Natl Acad Sci USA* 1996;93:295–299.

[294] Brette F, Rodriguez P, Komukai K, Colyer J, Orchard CH. beta-adrenergic stimulation restores the Ca2þ transient of ventricular myocytes lacking t-tubules. *J Mol Cell Cardiol* 2004;36:265–275.

[295] Brette F, Despa S, Bers DM, Orchard CH. Spatiotemporal characteristics of SR Ca^{2+} uptake and release in detubulated rat ventricular myocytes. *J Mol Cell Cardiol* 2005;39:804–812.

[296] Brette F, Orchard CH. No apparent requirement for neuronal sodium channels in excitation-contraction coupling in rat ventricular myocytes. *Circ Res* 2006;98:667–674.

[297] Komukai K, Brette F, Yamanushi TT, Orchard CH. K^+ current distribution in rat sub-epicardial ventricular myocytes. *Pflugers Arch* 2002;444: 532–538.

[298] Jorgensen AO, Jones LR. Immunoelectron microscopical localization of phospholamban in adult canine ventricular muscle. *J Cell Biol* 1987; 104:1343–1352.

[299] Greene AL, Lalli MJ, Ji Y, Babu GJ, Grupp I, Sussman M, Periasamy M. Overexpression of SERCA2b in the heart leads to an increase in sarcoplasmic reticulum calcium transport function and increased cardiac contractility. *J Biol Chem* 2000;275:24722–24727.

[300] Litwin SE, Li J, Bridge JH. Na-Ca exchange and the trigger for sarcoplasmic reticulum Ca^{2+} release: studies in adult rabbit ventricular myocytes. *Biophys J* 1998;75:359–371.

[301] Bers DM, Barry WH, Despa S. Intracellular Naþ regulation in cardiac myocytes. *Cardiovasc Res* 2003;57:897–912.

[302] Doroudgar S, Glembotski CC, Newconcepts of endoplasmic reticulumfunction in the heart: Programmed to conserve, *J Mol Cell Cardiol* (2012), http://dx. doi. org/10. 1016/j. yjmcc. 2012. 10. 006

[303] Simmen T, Lynes EM, Gesson K, Thomas G. Oxidative protein folding in the endoplasmic reticulum: tight links to the mitochondria-associated membrane (MAM). *Biochim Biophys Acta* 2010;1798:1465-73.

[304] Palade GE, Siekevitz P. Livermicrosomes; an integrated morphological and biochemical study. *J Biophys Biochem Cytol* 1956;2:171-200.

[305] Porter KR, Kallman FL. Significance of cell particulates as seen by electron microscopy. *Ann N Y Acad Sci* 1952;54:882-91.

[306] Doroudgar S, Glembotski CC. The cardiokine story unfolds: ischemic stress-induced protein secretion in the heart. *Trends Mol Med* 2011;17:207-14.

[307] Caro LG, Palade GE. Protein synthesis, storage, and discharge in the pancreatic exocrine cell. an autoradiographic study. *J Cell Biol* 1964;20:473-95.

[308] Lingappa VR, Blobel G. Early events in the biosynthesis of secretory and membrane proteins: the signal hypothesis. *Recent Prog Horm Res* 1980;36:451-75.

[309] Bers DM. Cardiac excitation–contraction coupling. *Nature* 2002; 415:198-205.

[310] Sobie EA, Lederer WJ. Dynamic local changes in sarcoplasmic reticulum calcium: physiological and pathophysiological roles. *J Mol Cell Cardiol* 2012;52:304-11.

[311] Mesaeli N, Nakamura K, Opas M, Michalak M. Endoplasmic reticulum in the heart, a forgotten organelle? *Mol Cell Biochem* 2001;225:1-6.

[312] Michalak M, Opas M. Endoplasmic and sarcoplasmic reticulum in the heart. *Trends Cell Biol* 2009;19:253-9.

[313] Belmont PJ, Tadimalla A, ChenWJ, Martindale JJ, Thuerauf DJ, Marcinko M, et al. Coordination of growth and endoplasmic reticulum stress signaling by regulator of calcineurin 1 (RCAN1), a novel ATF6-inducible gene. *J Biol Chem* 2008;283:14012-21.

[314] Kaisto T, Metsikko K. Distribution of the endoplasmic reticulum and its relationship with the sarcoplasmic reticulumin skeletal myofibers. *Exp Cell Res* 2003;289:47-57.

[315] Thuerauf DJ, Marcinko M, Gude N, Rubio M, SussmanMA, Glembotski CC. Activation of the unfolded protein response in infarcted mouse heart and hypoxic cultured cardiac myocytes. *Circ Res* 2006;99:275-82.

[316] Volpe P, Villa A, Podini P, Martini A, Nori A, Panzeri MC, et al. The endoplasmic reticulum–sarcoplasmic reticulum connection: distribution of endoplasmic reticulum markers in the sarcoplasmic reticulum of skeletal muscle fibers. *Proc Natl Acad Sci U S A* 1992;89:6142-6.

[317] Wu X, Zhang T, Bossuyt J, Li X, McKinsey TA, Dedman JR, et al. Local InsP3-dependent perinuclear Ca2+ signaling in cardiac myocyte excitation–transcription coupling. *J Clin Invest* 2006;116:675-82.

[318] Knollmann BC. A "rough" journey to the sarcoplasmic reticulum—implications of altered calsequestrin trafficking for cardiac arrhythmia. *JMol Cell Cardiol* 2010;49:554-5.

[319] Gerdes A. Cardiomyocyte ultrastructure. Muscle: In: Hill JA, Olsen EN, editors. Fundamental Biology and Mechanisms of Disease. San Diego: Academic Press; 2012. p. 47-55.

[320] McFarland TP, Milstein ML, Cala SE. Rough endoplasmic reticulum to junctional sarcoplasmic reticulum trafficking of calsequestrin in adult cardiomyocytes. *J Mol Cell Cardiol* 2010;49:556-64.

[321] Vangheluwe P, Louch WE, Ver Heyen M, Sipido K, Raeymaekers L,Wuytack F. Ca2+ transport ATPase isoforms SERCA2a and SERCA2b are targeted to the same sites in the murine heart. *Cell Calcium* 2003;34:457-64.

[322] Anfinsen CB. Principles that govern the folding of protein chains. *Science* 1973;181: 223-30.

[323] Hetz C. The unfolded protein response: controlling cell fate decisions under ER stress and beyond. *Nat Rev Mol Cell Biol* 2012;13:89–102.

[324] Walter P, Ron D. The unfolded protein response: fromstress pathway to homeostatic regulation. *Science* 2011;334:1081-6.

[325] Rutkowski DT, Hegde RS. Regulation of basal cellular physiology by the homeostatic unfolded protein response. *J Cell Biol* 2010;189:783-94.

[326] Glembotski CC. Endoplasmic reticulum stress in the heart. *Circ Res* 2007;101: 975-84.

[327] Glembotski CC. The role of the unfolded protein response in the heart. *J Mol Cell Cardiol* 2008;44:453-9.

[328] Minamino T, Komuro I, Kitakaze M. Endoplasmic reticulum stress as a therapeutic target in cardiovascular disease. *Circ Res* 2010;107:1071-82.

[329] Lai E, Teodoro T, Volchuk A. Endoplasmic reticulum stress: signaling the unfolded protein response. *Physiology* (Bethesda) 2007;22:193-201.

[330] Doroudgar S, Thuerauf DJ, Marcinko MC, Belmont PJ, Glembotski CC. Ischemia activates the ATF6 branch of the endoplasmic reticulum stress response. *J Biol Chem* 2009;284:29735-45.

[331] Groenendyk J, Sreenivasaiah PK, Kim do H, Agellon LB, Michalak M. Biology of endoplasmic reticulum stress in the heart. *Circ Res* 2010;107:1185-97.

[332] Harding HP, Zhang Y, Ron D. Protein translation and folding are coupled by an endoplasmic-reticulum-resident kinase. *Nature* 1999;397:271-4.

[333] Shi Y, Vattem KM, Sood R, An J, Liang J, Stramm L, et al. Identification and characterization of pancreatic eukaryotic initiation factor 2 alpha-subunit kinase, PEK, involved in translational control. *Mol Cell Biol* 1998;18:7499-509.

[334] Cox JS, Shamu CE,Walter P. Transcriptional induction of genes encoding endoplasmic reticulum resident proteins requires a transmembrane protein kinase. *Cell* 1993;73: 1197-206.

[335] TirasophonW, Welihinda AA, Kaufman RJ. A stress response pathway from the endoplasmic reticulum to the nucleus requires a novel bifunctional protein kinase/ endoribonuclease (Ire1p) in mammalian cells. *Genes Dev* 1998;12:1812-24.

[336] Wang XZ, Harding HP, Zhang Y, Jolicoeur EM, KurodaM, Ron D. Cloning ofmammalian Ire1 reveals diversity in the ER stress responses. *EMBO J* 1998;17:5708-17.

[337] Niwa M, Sidrauski C, Kaufman RJ, Walter P. A role for presenilin-1 in nuclear accumulation of Ire1 fragments and induction of the mammalian unfolded protein response. *Cell* 1999;99:691-702.

[338] Zhu C, Johansen FE, Prywes R. Interaction of ATF6 and serum response factor. *Mol Cell Biol* 1997;17:4957-66.

[339] Yoshida H, Haze K, Yanagi H, Yura T,Mori K. Identification of the cis-acting endoplasmic reticulum stress response element responsible for transcriptional induction of mammalian glucose-regulated proteins. Involvement of basic leucine zipper transcription factors. *J Biol Chem* 1998;273:33741-9.

[340] Haze K, Yoshida H, Yanagi H, Yura T, Mori K. Mammalian transcription factor ATF6 is synthesized as a transmembrane protein and activated by proteolysis in response to endoplasmic reticulum stress. *Mol Biol Cell* 1999;10:3787-99.

[341] Kilberg MS, Shan J, Su N. ATF4-dependent transcription mediates signaling of amino acid limitation. *Trends Endocrinol Metab* 2009;20:436-43.

[342] Plemper RK, Wolf DH. Endoplasmic reticulum degradation. Reverse protein transport and its end in the proteasome. *Mol Biol Rep* 1999;26:125-30.

[343] Clarke R, Cook KL, Hu R, Facey CO, Tavassoly I, Schwartz JL, et al. Endoplasmic reticulum stress, the unfolded protein response, autophagy, and the integrated regulation of breast cancer cell fate. *Cancer Res* 2012;72:1321-31.

[344] Bernales S, Schuck S, Walter P. ER-phagy: selective autophagy of the endoplasmic reticulum. *Autophagy* 2007;3:285-7.

[345] Bernales S, McDonald KL,Walter P. Autophagy counterbalances endoplasmic reticulum expansion during the unfolded protein response. *PLoS Biol* 2006;4:e423.

[346] Gustafsson AB, Gottlieb RA. Recycle or die: the role of autophagy in cardioprotection. *J Mol Cell Cardiol* 2008;44:654-61.

[347] Xu C, Bailly-Maitre B, Reed JC. Endoplasmic reticulum stress: cell life and death decisions. *J Clin Invest* 2005;115:2656-64.

[348] Thuerauf DJ, Morrison LE, Hoover H, Glembotski CC. Coordination of ATF6-mediated transcription and ATF6 degradation by a domain that is shared with the viral transcription factor, VP16. *J Biol Chem* 2002;277:20734-9.

[349] Okada K, Minamino T, Tsukamoto Y, Liao Y, Tsukamoto O, Takashima S, et al. Prolonged endoplasmic reticulum stress in hypertrophic and

failing heart after aortic constriction: possible contribution of endoplasmic reticulumstress to cardiac myocyte apoptosis. *Circulation* 2004;110:705-12.

[350] Thuerauf DJ, Hoover H, Meller J, Hernandez J, Su L, Andrews C, et al. Sarco/endoplasmic reticulum calcium ATPase-2 expression is regulated by ATF6 during the endoplasmic reticulum stress response: intracellular signaling of calcium stress in a cardiac myocyte model system. *J Biol Chem* 2001;276:48309-17.

[351] Terai K, Hiramoto Y, Masaki M, Sugiyama S, Kuroda T, Hori M, et al. AMP-activated protein kinase protects cardiomyocytes against hypoxic injury through attenuation of endoplasmic reticulum stress. *Mol Cell Biol* 2005;25:9554-75.

[352] Bousette N, Chugh S, Fong V, Isserlin R, KimKH, Volchuk A, et al. Constitutively active calcineurin induces cardiac endoplasmic reticulum stress and protects against apoptosis that is mediated by alpha-crystallin-B. *Proc Natl Acad Sci U S A* 2010;107: 18481-6.

[353] Martindale JJ, Fernandez R, Thuerauf D, Whittaker R, Gude N, Sussman MA, et al. Endoplasmic reticulumstress gene induction and protection fromischemia/reperfusion injury in the hearts of transgenic mice with a tamoxifen-regulated form of ATF6. *Circ Res* 2006;98:1186-93.

[354] Toko H, Takahashi H, Kayama Y, Okada S, Minamino T, Terasaki F, et al. ATF6 is important under both pathological and physiological states in the heart. *J Mol Cell Cardiol* 2010;49:113-20.

[355] Lynch JM, Maillet M, Vanhoutte D, Schloemer A, Sargent MA, Blair NS, et al. A thrombospondin-dependent pathway for a protective ER stress response. *Cell* 2012;149:1257-68.

[356] Belmont PJ, Chen WJ, Thuerauf DJ, Glembotski CC. Regulation of microRNA expression in the heart by the ATF6 branch of the ER stress response. *J Mol Cell Cardiol* 2012;52:1176-82.

[357] Kitakaze M, Tsukamoto O. What is the role of ER stress in the heart? Introduction and series overview. *Circ Res* 2010;107:15-8.

[358] Ni L, Zhou C, Duan Q, Lv J, Fu X, Xia Y, et al. Beta-AR blockers suppresses ER stress in cardiac hypertrophy and heart failure. *PLoS One* 2011;6:e27294.

[359] Dalal S, Foster CR, Das BC, Singh M, Singh K. Beta-adrenergic receptor stimulation induces endoplasmic reticulum stress in adult cardiac myocytes: role in apoptosis. *Mol Cell Biochem* 2012;364:59-70.

[360] Qi X, Vallentin A, Churchill E,Mochly-RosenD. deltaPKC participates in the endoplasmic reticulumstress-induced response in cultured cardiacmyocytes and ischemic heart. *J Mol Cell Cardiol* 2007;43:420-8.
[361] Szegezdi E, Duffy A, O'Mahoney ME, Logue SE, Mylotte LA, O'Brien T, et al. ER stress contributes to ischemia-induced cardiomyocyte apoptosis. *Biochem Biophys Res Commun* 2006;349:1406-11.
[362] Bravo R, Gutierrez T, Paredes F, Gatica D, Rodriguez AE, Pedrozo Z, et al. Endoplasmic reticulum: ER stress regulates mitochondrial bioenergetics. *Int J Biochem Cell Biol* 2012;44:16-20.
[363] Rizzuto R, Brini M, Murgia M, Pozzan T. Microdomains with high Ca^{2+} close to IP3-sensitive channels that are sensed by neighboring mitochondria. *Science* 1993;262:744-7.
[364] Giacomello M, Drago I, Bortolozzi M, Scorzeto M, Gianelle A, Pizzo P, et al. Ca^{2+} hot spots on the mitochondrial surface are generated by Ca2+ mobilization from stores, but not by activation of store-operated Ca2+ channels. *Mol Cell* 2010;38:280-90.
[365] Csordas G, Varnai P, Golenar T, Roy S, Purkins G, Schneider TG, et al. Imaging interorganelle contacts and local calcium dynamics at the ER–mitochondrial interface. *Mol Cell* 2010;39:121-32.
[366] Csordas G, Renken C, Varnai P, Walter L, Weaver D, Buttle KF, et al. Structural and functional features and significance of the physical linkage between ER and mitochondria. *J Cell Biol* 2006;174:915-21.
[367] Hayashi T, Rizzuto R, Hajnoczky G, Su TP. MAM: more than just a housekeeper. *Trends Cell Biol* 2009;19:81-8.
[368] Raturi A, Simmen T. Where the endoplasmic reticulumand themitochondrion tie the knot: the mitochondria-associated membrane (MAM). Biochim Biophys Acta 2012, http://dx. doi. org/10. 1016/j. bbamcr. 2012. 04. 013.
[369] Marks AR. Intracellular calcium-release channels: regulators of cell life and death. *Am J Physiol* 1997;272:H597-605.
[370] de BritoOM, Scorrano L. Mitofusin 2 tethers endoplasmic reticulumto mitochondria. *Nature* 2008;456:605-10.
[371] De Stefani D, Raffaello A, Teardo E, Szabo I, Rizzuto R. A forty-kilodalton protein of the inner membrane is the mitochondrial calcium uniporter. *Nature* 2011;476:336-40.
[372] Szabadkai G, DuchenMR. Mitochondria: the hub of cellular Ca2+ signaling. *Physiology* (Bethesda) 2008;23:84-94.

[373] Garcia-Perez C, Schneider TG, Hajnoczky G, Csordas G. Alignment of sarcoplasmic reticulum-mitochondrial junctions with mitochondrial contact points. *Am J Physiol Heart Circ Physiol* 2011;301:H1907-15.

[374] Papanicolaou KN, Khairallah RJ, Ngoh GA, Chikando A, Luptak I, O'Shea KM, et al. Mitofusin-2 maintains mitochondrial structure and contributes to stress-induced permeability transition in cardiac myocytes. *Mol Cell Biol* 2011;31:1309-28.

[375] Chen Y, Csordas G, Jowdy C, Schneider TG, Csordas N, Wang W, et al. Mitofusin 2-containing mitochondrial-reticular microdomains direct rapid cardiomyocyte bioenergetics responses via inter-organelle Ca2+ crosstalk. *Circ Res* 2012;111:863-75.

[376] Ngoh GA, Papanicolaou KN, Walsh K. Loss of mitofusin 2 promotes endoplasmic reticulum stress. *J Biol Chem* 2012;287:20321-32.

[377] Kumar S, Kasseckert S, Kostin S, Abdallah Y, Schafer C, Kaminski A, et al. Ischemic acidosis causes apoptosis in coronary endothelial cells through activation of caspase-12. *Cardiovasc Res* 2007;73:172-80.

[378] Valverde CA, Kornyeyev D, FerreiroM, Petrosky AD,Mattiazzi A, Escobar AL. Transient Ca2+ depletion of the sarcoplasmic reticulum at the onset of reperfusion. *Cardiovasc Res* 2010;85:671-80.

[379] Piper HM, Abdallah Y, Schafer C. The first minutes of reperfusion: a window of opportunity for cardioprotection. *Cardiovasc Res* 2004;61:365-71.

[380] Kusuoka H, Porterfield JK, Weisman HF, Weisfeldt ML, Marban E. Pathophysiology and pathogenesis of stunnedmyocardium. Depressed Ca2+ activation of contraction as a consequence of reperfusion-induced cellular calciumoverload in ferret hearts. *J Clin Invest* 1987;79:950-61.

[381] Nicchitta CV, Lerner RS, Stephens SB, Dodd RD, Pyhtila B. Pathways for compartmentalizing protein synthesis in eukaryotic cells: the template-partitioning model. *Biochem Cell Biol* 2005;83:687-95.

[382] Reid DW, Nicchitta CV. Primary role for endoplasmic reticulum-bound ribosomes in cellular translation identified by ribosome profiling. *J Biol Chem* 2012;287: 5518-27.

[383] Foster KG, Fingar DC. Mammalian target of rapamycin (mTOR): conducting the cellular signaling symphony. *J Biol Chem* 2010; 285:14071-7.

[384] Proud CG. Ras, PI3-kinase and mTOR signaling in cardiac hypertrophy. *Cardiovasc Res* 2004;63:403-13.

[385] Hedhli N, Pelat M, Depre C. Protein turnover in cardiac cell growth and survival. *Cardiovasc Res* 2005;68:186-96.

[386] Aoyagi T, Matsui T. Phosphoinositide-3 kinase signaling in cardiac hypertrophy and heart failure. *Curr Pharm Des* 2011;17:1818-24.

[387] Appenzeller-Herzog C, Hall MN. Bidirectional crosstalk between endoplasmic reticulum stress and mTOR signaling. *Trends Cell Biol* 2012;22:274-82.

[388] Zoncu R, Bar-Peled L, Efeyan A, Wang S, Sancak Y, Sabatini DM. mTORC1 senses lysosomal amino acids through an inside-out mechanism that requires the vacuolar H(+)-ATPase. *Science* 2011;334:678-83.

[389] Zinzalla V, Stracka D, Oppliger W, Hall MN. Activation of mTORC2 by association with the ribosome. *Cel*l 2011;144:757-68.

[390] Drenan RM, Liu X, Bertram PG, Zheng XF. FKBP12-rapamycin-associated protein or mammalian target of rapamycin (FRAP/mTOR) localization in the endoplasmic reticulum and the Golgi apparatus. *J Biol* Chem 2004;279:772-8.

[391] Boulbes DR, Shaiken T, Sarbassov dos D. Endoplasmic reticulum is a main localization site of mTORC2. *Biochem Biophys Res Commun* 2011;413:46-52.

[392] Liu X, Zheng XF. Endoplasmic reticulum and Golgi localization sequences for mammalian target of rapamycin. *Mol Biol Cell* 2007;18:1073-82.

[393] Deldicque L, Hespel P, Francaux M (2011) ER stress in skeletal muscle: origin and metabolic consequences. *Exerc Sport Sci Rev*.

[394] Dirksen RT Sarcoplasmic Reticulum-Mitochondrial "Through-Space" Coupling in Skeletal Muscle. *Appl Physiol Nutr Metab.* 2009 Jun; 34(3): 389–395.

[395] Rossi AE, Boncompagni S, Dirksen RT. Sarcoplasmic reticulum-mitochondrial symbiosis: bidirectional signaling in skeletal muscle. *Exerc. Sport Sci. Rev*. 2009;37:29–35.

[396] Wang X, Weisleder N, Collet C, Zhou J, Chu Y, Hirata Y, Zhao X, Pan Z, Brotto M, Cheng H, Ma J. Uncontrolled calcium sparks act as a dystrophic signal for mammalian skeletal muscle. *Nat. Cell. Biol.* 2005;7:525–530.

[397] Weisleder N, Brotto M, Komazaki S, Pan Z, Zhao X, Nosek T, Parness J, Takeshima H, Ma J. Muscle aging is associated with compromised Ca2+ spark signaling and segregated intracellular Ca2+ release. *J. Cell* Biol. 2006;174:639–645.

[398] Durham WJ, Aracena-Parks P, Long C, Rossi AE, Goonasekera SA, Boncompagni S, Galvan DL, Gilman CP, Baker MR, Shirokova N,

Protasi F, Dirksen R, Hamilton SL. RyR1 S-nitrosylation underlies environmental heat stroke and sudden death in Y522S RyR1 knockin mice. *Cell.* 2008;133:53–65.

[399] Gabella G. Caveolae intracellulaires and sarcoplasmic reiculum in smooth muscle. *J Cell Sci* 8: 601–609, 1971.

[400] Gabella G. Structure of smooth muscle. In: Smooth Muscle: An Assessment of Current Knowledge, edited by Bulbring E, Brading AF, Jones AW and Tomita T. London: Arnold, 1981.

[401] Gabella G. Structural apparatus for force transmission in smooth muscles. *Physiol Rev* 64: 455–477, 1984.

[402] Somlyo AV, Somlyo AP. Strontium accumulation by sarcoplasmic reticulum and mitochondria in vascular smooth muscle. *Science* 174: 955–958, 1971.

[403] Butler TM, Davies RE. High-energy phosphates in smooth muscle. In: Handbook of Physiology. The Cardiovascular System. Vascular Smooth Muscle. Bethesda, MD: *Am. Physiol. Soc.*, 1980, sect. 2, vol. II, chapt. 10, p. 237–252.

[404] Kowarski D, Shutman H, Somlyo AP, Somlyo AV. Calcium release by noradrenaline from central sarcoplasmic reticulum in rabbit pulmonary artery smooth muscle. *J Physiol* 366: 153–175, 1985.

[405] Bond M, Kitazawa T, Somlyo AP, Somlyo AV. Release and recycling of calcium by the sarcoplasmic reticulum in guinea-pig portal vein smooth muscle. *J Physiol* 355: 677–695, 1984.

[406] Bonev AD, Jagger JH, Rubart M, Nelson MT. Activators of protein kinase C decreases Ca2 spark frequency in smooth muscle cells from cerebral arteries. *Am J Physiol Cell Physiol* 273: C2090–C2095, 1997.

[407] Jaggar JH, Wellman GC, Heppner TJ, Porter VA, Perez GJ, Gollasch M, Kleppisch T, Rubart M, Stevenson AS, Lederer WJ, Knot HJ, Bonev AD, Nelson MT. Ca2 channels, ryanodine receptors and Ca2-activated K channels: a functional unit for regulating arterial tone. *Acta Physiol Scand* 164: 577–587, 1998.

[408] Nelson MT, Cheng H, Rubart M, Santana LF, Bonev AD, Knot HJ, Lederer WJ. Relaxation of arterial smooth muscle by calcium sparks. *Science* 270: 633–637, 1995.

[409] Borisova L, Shmygol A, Wray S, Burdyga T. Evidence that a Ca2 sparks/STOCs coupling mechanism is responsible for the inhibitory effect of caffeine on electro-mechanical coupling in guinea pig ureteric smooth muscle. *Cell Calcium* 42: 303–311, 2007.

[410] Taggart MJ, Wray S. Contribution of sarcoplasmic reticular calcium to smooth muscle contractile activation: gestational dependence in isolated rat uterus. *J Physiol* 511: 133–144, 1998.

[411] Burdyga T, Wray S. Action potential refractory period in ureter smooth muscle is set by Ca sparks and BK channels. *Nature* 436: 559–562, 2005.

[412] Chamley-Campbell J, Campbell GR, Ross R. The smooth muscle cell in culture. *Physiol Rev* 59: 1–61, 1979.

[413] Owens GK. Regulation of differentiation of vascular smooth muscle cells. *Physiol Rev* 75: 487–517, 1995.

[414] Holly A. Shiels and Gina L. J. *Galli The Sarcoplasmic Reticulum and the Evolution of the Vertebrate Heart Physiology* 29:456-469, 2014.

[415] Galli GJ, Shiels H. The sarcoplasmic reticulum in the vertebrate heart. In: Ontogeny and Phylogeny of the Vertebrate Heart, edited by Sedmera D, Wang T. New York: Springer, 2012, p. 103–124.

[416] Tibbits GF, Hovemadsen L, Bers DM. Calciumtransport and the regulation of cardiac contractility in teleosts: a comparison with higher vertebrates. *Can J Zool* 69: 2014–2019, 1991.

[417] Vornanen M, Shiels HA, Farrell AP. Plasticity of excitation-contraction coupling in fish cardiac myocytes. *Comp Biochem Physiol* 132: 827–846, 2002.

[418] Bers DM. Ca2influx and sarcoplasmic reticulum Ca release in cardiac muscle activation during postrest recovery. *Am J Physiol Heart Circ Physiol* 248: H366–H381, 1985.

[419] Bers DM. Excitation-Contraction Coupling and Contractile Force. Dordecht, The Netherlands: Kluwer Academic, 2001.

[420] Bers DM, Bridge JH. Relaxation of rabbit ventricular muscle by Na-Ca exchange and sarcoplasmic reticulum calcium pump. Ryanodine and voltage sensitivity. *Circ Res* 65: 334–342, 1989.

[421] Haddock PS, Coetzee WA, Cho E, Porter L, Katoh H, Bers DM, Jafri MS, Artman M. Subcellular [Ca2]i gradients during excitation-contraction coupling in newborn rabbit ventricular myocytes. *Circ Res* 85: 415–427, 1999.

[422] Galli GLJ. The role of the sarcoplasmic reticulum in the generation of high heart rates and blood pressures in reptiles. *J Exp Biol* 209: 1956–1963, 2006.

[423] Ravens U, Gath J, Hussaini MA, Himmel H. Mechanical restitution in atrial muscle from human and rat hearts: effects of agents that modify sarcoplasmic reticulum function. *Pharm Toxicol* 81: 97–104, 1997.

[424] Tanaka H, Takagi N, Shigenobu K. Difference in excitation-contraction mechanisms between atrial and ventricular myocardia of hatched chicks. *Gen Pharm* 26: 45–49, 1995.

[425] Shiels HA, Farrell AP. The effect of temperature and adrenaline on the relative importance of the sarcoplasmic reticulum in contributing Ca2 to force development in isolated ventricular trabeculae from rainbow trout. *J Exp Biol* 200: 1607–1621, 1997.

[426] Shiels HA, Freund EV, Farrell AP, Block BA. The sarcoplasmic reticulum plays a major role in isometric contraction in atrial muscle of yellowfin tuna. *J Exp Biol* 202: 881–890, 1999.

[427] Galli GL, Lipnick MS, Shiels HA, Block BA. Temperature effects on Ca2cycling in scombrid cardiomyocytes: a phylogenetic comparison. *J Exp Biol* 214: 1068–1076, 2011.

[428] Haverinen J, Vornanen M. Comparison of sarcoplasmic reticulum calcium content in atrial and ventricular myocytes of three fish species. *Am J Physiol Regul Integr Comp Physiol* 297: R1180–R1187, 2009.

[429] Bootman MD, Higazi DR, Coombes S, Roderick HL. Calcium signalling during excitation-contraction coupling in mammalian atrial myocytes. *J Cell Sci* 119: 3915–3925, 2006.

[430] Fabiato A. Calcium release in skinned cardiac cells: variations with species, tissues, and development. *Fed Proc* 41: 2238–2244, 1982. 432.

[431] Franzini-Armstrong C, Protasi F, Tijskens (2005) P. The assembly of calcium release units in cardiac muscle. *Ann NY Acad Sci* 1047: 76–85.

[432] Vornanen M, Shiels HA, Farrell AP. Plasticity of excitation-contraction coupling in fish cardiac myocytes. *Comp Biochem Physiol* 132: 827–846, 2002.

[433] Stern MD, Pizarro G, Ríos E. Local control model of excitation-contraction coupling in skeletal muscle. *J Gen Physiol* 110: 415–440, 1997.

[434] Stern MD. Theory of excitation-contraction coupling in cardiac muscle. *Biophys J* 63: 497–517, 1992.

[435] Satoh H, Delbridge LM, Blatter LA, Bers DM. Surface:volume relationship in cardiac myocytes studied with confocal microscopy and membrane capacitance measurements: species-dependence and developmental effects. *Biophys J* 70: 1494–1504, 1996.

[436] Vornanen M. L-type Ca2 current in fish cardiac myocytes: effects of thermal acclimation and beta-adrenergic stimulation. *J Exp Biol* 201: 533–547, 1998.

[437] Galli GLJ. The role of the sarcoplasmic reticulum in the generation of high heart rates and blood pressures in reptiles. *J Exp Biol* 209: 1956–1963, 2006.

[438] Kim CS, Davidoff AJ, Maki TM, Doye AA, Gwathmey JK. Intracellular calcium and the relationship to contractility in an avian model of heart failure. *J Comp Physiol A* 170: 295–306, 2000.

[439] Goaillard JM, Vincent PV, Fischmeister R. Simultaneous measurements of intracellular cAMP and L-type Ca2current in single frog ventricular myocytes. *J Physiol* 530: 79–91, 2001.

[440] Bean BP, Nowycky MC, Tsien RW. β-Adrenergic modulation of calcium channels in frog ventricular heart cells. *Nature* 307: 371–375, 1984.

[441] Shiels HA, White E. Temporal and spatial properties of cellular Ca^{2+} flux in trout ventricular myocytes. *Am J Physiol Regul Integr Comp Physiol* 288: R1756–R1766, 2005.

[442] Jewett PH, Sommer JR, Johnson EA. Cardiac muscle. Its ultrastructure in the finch and hummingbird with special reference to the sarcoplasmic reticulum. *J Cell Biol* 49: 50–65, 1971.

[443] Brette F, Luxan G, Cros C, Dixey H, Wilson C, Shiels HA. Characterization of isolated ventricular myocytes from adult zebrafish (Danio rerio). *Biochem Biophys Res Commun* 374: 143–146, 2008.

[444] Vornanen M, Haverinen J. A significant role of sarcoplasmic reticulum in cardiac contraction of a basal vertebrate, the river lamprey (*Lampetra fluviatilis*). *Acta Physiol* (Oxf) 207: 269–279, 2013.

[445] Pinali C, Bennett H, Davenport JB, Trafford AW, Kitmitto A. Three-dimensional reconstruction of cardiac sarcoplasmic reticulum reveals a continuous network linking transverse-tubules: this organization is perturbed in heart failure. *Circ Res* 113: 1219–1230, 2013.

[446] Eisner DA. Sodium calcium exchange in the heart: necessity or luxury? *Circ Res* 95: 549–551, 2004.

[447] Bootman MD, Higazi DR, Coombes S, Roderick HL. Calcium signalling during excitation-contraction coupling in mammalian atrial myocytes. *J Cell Sci* 119: 3915–3925, 2006.

[448] Perni S, Iyer VR, Franzini-Armstrong C. Ultrastructure of cardiac muscle in reptiles and birds: optimizing and/or reducing the probability of transmission between calcium release units. *J Muscle Res Cell Motil* 33: 145–152, 2012.

[449] Cheng H, Lederer WJ, Cannell MB. Calcium sparks: elementary events underlying excitationcontraction coupling in heart muscle. *Science* 262: 740–744, 1993.

[450] Cannell MB, Cheng H, Lederer WJ. The control of calcium release in heart muscle. *Science* 268: 1045–1049, 1995.

[451] Mackenzie L, Roderick HL, Berridge MJ, Conway SJ, Bootman MD. The spatial pattern of atrial cardiomyocyte calcium signalling modulates contraction. *J Cell Sci* 117: 6327–6337, 2004.

[452] Pinali C, Bennett H, Davenport JB, Trafford AW, Kitmitto A. Three-dimensional reconstruction of cardiac sarcoplasmic reticulum reveals a continuous network linking transverse-tubules: this organization is perturbed in heart failure. *Circ Res* 113: 1219–1230, 2013.

[453] Kirk MM, Izu LT, Chen-Izu Y, McCulle SL, Wier WG, Balke CW, Shorofsky SR. Role of the transverse- axial tubule system in generating calcium sparks and calcium transients in rat atrial myocytes. *J Physiol* 547: 441–451, 2003.

[454] Richards MA, Clarke JD, Saravanan P, Voigt N, Dobrev D, Eisner DA, Trafford AW, Dibb KM. Transverse tubules are a common feature in large mammalian atrial myocytes including human. *Am J Physiol Heart Circ Physiol* 301: H1996–H2005, 2011.

[455] Baddeley D, Jayasinghe ID, Lam L, Rossberger S, Cannell MB, Soeller C. Optical single-channel resolution imaging of the ryanodine receptor distribution in rat cardiac myocytes. *Proc Natl Acad Sci USA* 106: 22275–22280, 2009.

[456] Chen-Izu Y, McCulle SL, Ward CW, Soeller C, Allen BM, Rabang C, Cannell MB, Balke CW, Izu LT. Three-dimensional distribution of ryanodine receptor clusters in cardiac myocytes. *Biophys J* 91: 1–13, 2006.

[457] Bers DM, Shannon TR. Calcium movements inside the sarcoplasmic reticulum of cardiac myocytes. *J Mol Cell Cardiol* 58: 59–66, 2013.

[458] Niggli E, Ullrich ND, Gutierrez D, Kyrychenko S, Poláková E, Shirokova N. Posttranslational modifications of cardiac ryanodine receptors: Ca2 signaling and EC-coupling. *Biochim Biophys Acta* 1833: 866–875, 2013.

[459] Santana LF, Cheng H, Gomez AM, Cannell MB, Lederer WJ. Relation between the sarcolemmal Ca^{2+} current and Ca^{2+} sparks and local control theories for cardiac excitation-contraction coupling. *Circ Res* 78: 166–171, 1996.

[460] Marx SO, Gaburjakova J, Gaburjakova M, Henrikson C, Ondrias K, Marks AR. Coupled gating between cardiac calcium release channels (ryanodine receptors). *Circ Res* 88: 1151–1158, 2001.

[461] Marx SO, Reiken S, Hisamatsu Y, Jayaraman T, Burkhoff D, Rosemblit N, Marks AR. PKA phosphorylation dissociates FKBP12. 6 from the calcium release channel (ryanodine receptor): defective regulation in failing hearts. *Cell* 101: 365–376, 2000.

[462] Park H, Park IY, Kim E, Youn B, Fields K, Dunker AK, Kang C. Comparing skeletal and cardiac calsequestrin structures and their calcium binding: a proposed mechanism for coupled calcium binding and protein polymerization. *J Biol Chem* 279: 18026–18033, 2004.

[463] Vornanen M, Shiels HA, Farrell AP. Plasticity of excitation-contraction coupling in fish cardiac myocytes. *Comp Biochem Physiol* 132: 827–846, 2002.

[464] Tiitu V, Vornanen M. Regulation of cardiac contractility in a stenothermal fish, the burbot (Lota lota). *J Exp Biol* 205: 1597–1606, 2002.

[465] Shiels HA, Di Maio A, Thompson S, Block BA. Warm fish with cold hearts: thermal plasticity of excitation-contraction coupling in bluefin tuna. *Proc Roy Soc B* 278: 18–27, 2011.

[466] Li XC, Wei L, Zhang GQ, Bai ZL, Hu YY, Zhou P, Bai SH, Chai Z, Lakatta EG, Hao XM, Wang SQ. Ca^{2+} cycling in heart cells from ground squirrels: adaptive strategies for intracellular Ca^{2+} homeostasis. *PLos One* 6: e24787, 2011.

[467] Bovo E, Dvornikov AV, Mazurek SR, de Tombe PP, Zima AV. Mechanisms of Ca^{2+} handling in zebrafish ventricular myocytes. *Pflügers Arch* 465: 1775–1784, 2013.

[468] Shiels HA, White E. Temporal and spatial properties of cellular Ca^{2+} flux in trout ventricular myocytes. *Am J Physiol Regul Integr Comp Physiol* 288: R1756–R1766, 2005.

[469] Klitzner T, Morad M. Excitation-contraction coupling in frog ventricle. *Pflügers Arch* 398: 274– 283, 1983.

[470] Vornanen M. Regulation of contractility of the fish (Carassius-Carassius L) heart ventricle. *Comp Biochem Physiol* 94: 477–483, 1989.

[471] Tiitu V, Vornanen M. Regulation of cardiac contractility in a stenothermal fish, the burbot (Lota lota). *J Exp Biol* 205: 1597–1606, 2002.

[472] Shiels HA, Di Maio A, Thompson S, Block BA. Warm fish with cold hearts: thermal plasticity of excitation-contraction coupling in bluefin tuna. *Proc Roy Soc B* 278: 18–27, 2011.

[473] Shiels HA, White E. Temporal and spatial properties of cellular Ca^{2+} flux in trout ventricular myocytes. *Am J Physiol Regul Integr Comp Physiol* 288: R1756–R1766, 2005.

[474] Bowler K, Tirri R. Temperature dependence of the heart isolated from the cold or warm acclimated perch (*Perca fluviatilis*). *Comp Biochem Physiol* 96: 177–180, 1990.

[475] Sutko JL, Airey JA. Ryanodine receptor Ca^{2+} release channels: does diversity in form equal diversity in function? *Physiol Rev* 76: 1027–1071, 1996.

[476] Vornanen M. Temperature and Ca^{2+} dependence of [H-3]ryanodine binding in the burbot (*Lota lota* L.) heart. *Am J Physiol Regul Integr Comp Physiol* 290: R345–R351, 2006.

[477] Korajoki H, Vornanen M. Expression of calsequestrin in atrial and ventricular muscle of thermally acclimated rainbow trout. *J Exp Biol* 212: 3403–3414, 2009.

[478] Korajoki H, Vornanen M. Expression of SERCA and phospholamban in rainbow trout (*Oncorhynchus mykiss*) heart: comparison of atrial and ventricular tissue and effects of thermal acclimation. *J Exp Biol* 215: 1162–1169, 2012.

[479] Greising SM, Gransee HM, Mantilla CB, Sieck GC Systems Biology of Skeletal Muscle: Fiber Type as an Organizing Principle. *Wiley Interdiscip Rev Syst Biol Med.* 2012 ; 4(5)doi:10. 1002/wsbm. 1184.

[480] Fawcett DW, McNutt NS. The ultrastructure of the cat myocardium. I. Ventricular papillary muscle. *J Cell Biol.* 1969; 42:1–45.

[481] Franzini-Armstrong C, Protasi F and RAMESH V. Comparative Ultrastructure of Ca^{2+} Release Units in Skeletal and Cardiac Muscle. *Annals New York Academy of Sciences*. 1999, 20-30.

[482] Tajika Y, Takahashi M, Ueno H, Murakami T, Yorifuji H. Organization of organelles and VAMP-associated vesicular transport systems in differentiating skeletal muscle cells. *Anat Sci Int* (2015) 90:33–39 DOI 10. 1007/s12565-014-0266-6.

[483] Flucher BE, Franzini-Armstrong C. Formation of junctions involved in excitation-contraction coupling in skeletal and cardiac muscle. *Proc. Natl. Acad. Sci. USA* 1996, Vol. 93, pp. 8101-8106.

[484] Brillantes AB, Bezprozvannaya S, Marks AR. Developmental and Tissue-Specific Regulation of Rabbit Skeletal and Cardiac Muscle

Calcium Channels Involved in Excitation-Contraction Coupling. Circulation Research 1994, 75:503-510. doi: 10. 1161/01. RES. 75. 3. 503.

[485] Rossi AE, Dirksen RT. Sarcoplasmic Reticulum: The Dynamic Calcium Governor of Muscle. *Muscle Nerve* 2006, 33: 715–731.

[486] Doroudgar S, Glembotski CC, Newconcepts of endoplasmic reticulumfunction in the heart: Programmed to conserve, J Mol Cell Cardiol (2012), http://dx. doi. org/10. 1016/j. yjmcc. 2012. 10. 006.

[487] Wray S, Burdyga Th. Sarcoplasmic Reticulum Function in Smooth Muscle. *Physiol Rev* 2010, 90: 113–178, doi:10. 1152/physrev. 00018. 2008.

Author's Contact Information

Dr. Rigers Bakiu
Associate Professor
Department of Aquaculture and Fisheries,
Faculty of Agriculture and Environment,
Agricultural University of Tirana
Lecturer/Researcher and Head,
Aquaculture and Fisheries Research Group
Email: rigers.bakiu@ubt.edu.al

Index

B

C

D

E

F

G

H

I

N

O

P

Q

R

S

T